Chronobiology of Marine Organisms

Do intertidal organisms simply respond to the rise and fall of tides, or do they possess biological timing and navigation mechanisms that allow them to anticipate when conditions are most favourable? How are the patterns of growth, development and reproduction of some marine plants and animals related to changes in day length or to phases of the moon? The author describes how marine organisms, from single cells to vertebrates, on seashores, in estuaries and in the open ocean, have evolved inbuilt biological clockwork and synchronization mechanisms that control rhythmic processes and navigational behaviour, permitting successful exploitation of highly variable and often hostile environments. Adopting a hypothesis-testing and experimental approach, the book is intended for undergraduate and postgraduate students of marine biology, marine ecology, animal behaviour, oceanography and other biological sciences and also as an introduction for researchers, including physiologists, biochemists and molecular biologists entering the field of chronobiology.

ERNEST NAYLOR is Professor Emeritus at the School of Ocean Sciences, Bangor University, where he was Lloyd Roberts Professor of Marine Zoology and Head of School of Ocean Sciences. He has published over 160 scientific publications and has been a Council Member of the UK Natural Environment Research Council, the UK Marine Biological Association, the Challenger Society for Marine Sciences, and a President of the Society for Experimental Biology and the Estuarine and Coastal Sciences Association. He has participated in a House of Lords Select Sub-Committee and various UK and European Commission co-ordinating committees for Marine Science and Technology. He has been involved in reviews of scientific programmes of several marine laboratories in the UK and France, and of UK-supported fisheries projects in the developing world. In 1998 he was appointed OBE. This is his third book.

Chronobiology of Marine Organisms

ERNEST NAYLOR

Professor Emeritus,
School of Ocean Sciences,
Bangor University, Wales

CAMBRIDGE UNIVERSITY PRESS

Cambridge, New York, Melbourne, Madrid, Cape Town, Singapore,
São Paulo, Delhi, Dubai, Tokyo

Cambridge University Press
The Edinburgh Building, Cambridge CB2 8RU, UK

Published in the United States of America by
Cambridge University Press, New York

www.cambridge.org
Information on this title: www.cambridge.org/9780521760539

First published 2010

Printed in the United Kingdom at the University Press, Cambridge

A catalogue record for this publication is available from the British Library

ISBN 978-0-521-76053-9 Hardback

For Gillian

Contents

The plates will be found between pages 86 and 87.

Preface

There is increasing recognition of chronobiology in our understanding of the time-base of ecology, behaviour and physiology of plants and animals. However, much of the scientific effort so far in this field of study has focussed on daily and seasonal rhythmicity associated with solar periodicity of the environment. Impressively, this has led to the concept of heritable circadian biological clocks and a search for their molecular basis in the genetic makeup of living systems. Partly because of early and perhaps lingering scepticism, the possibility that some organisms might also innately phase their behaviour to lunar events has lagged behind as a field of study. Yet, living organisms in many seas and coasts are repeatedly exposed to lunar cycles, indirectly through oscillations of ocean tides. Moreover, marine animals and plants have been in existence for greater lengths of evolutionary time than have the terrestrial organisms that are often the material for classical studies of circadian rhythmicity. It is therefore reasonable to consider the extent to which marine organisms have adapted to tidal oscillations driven by lunar gravity, and also to ask whether lunar and semilunar events exhibited by such organisms are related to fortnightly variations in tidal height or even to moonlight cycles directly. Accordingly, alongside the concept of circadian and circa-annual rhythms in marine organisms, it is necessary to consider the existence of innate biological clocks of circatidal, circasemilunar and circalunar periodicities.

Against the background of the physical basis of tides, and adopting a hypothesis-testing and experimental approach, this book explores the phenomena of biological rhythms and clocks in coastal, estuarine and open sea organisms in an ecological context. It considers the role of the diverse physical variables associated with tidal oscillations in synchronizing biological clocks of tide- and moon-related periodicities. It then assesses the relevance of innate biological timing capability to the cyclical changes of orientational and navigational behaviour of some marine animals, which permit them, after wide dispersal, to return to optimal zones on a seashore or to preferred locations in an estuary or in the open ocean, ensuring their occurrence in the right place at the right time. Finally, it outlines aspects of the search for the nature of biological clockwork, leading from the techniques of classical endocrinology to those of modern molecular biology. A fuller understanding of the molecular nature of circatidal clockwork and its relationship with better understood circadian molecular clock mechanisms remain challenges which curiosity-driven science has yet to resolve. It is hoped that this book will help to stimulate that scientific endeavour.

Several individuals have kindly read and commented on all or parts of drafts of the book, or have otherwise contributed help and discussion, namely: David Bowers, Ed Hill, Joanna Jones, Bambos Kiriacou, Irshad Mobarak, Elfed Morgan, Gillian Naylor, Graham Walker, Simon Webster and David Wilcockson. Many graduate students, post-doctorals and other collaborators, whose names appear as personal research associates in the text and bibliography, have also contributed greatly to the development of the subject of biorhythms in marine organisms as presented here, which I hope they recognize. At the Marine Science Laboratories, Menai Bridge, David Roberts has given considerable help in preparing the illustrations in an appropriate format, Graham Worley and Judy Davies gave excellent computer and secretarial assistance, respectively, and invaluable help was provided by the Woolfson Library staff. David Wilcockson generously prepared Figures 10.10 and 10.11 and Stephen Hanbury and John Hughes kindly provided Plates 9 and 14 respectively. Permission to reproduce the D. P. Wilson photograph (Plate 15) from A. C. Hardy (1956), *The Open Sea: World of Plankton*, 335 pp., Collins, London was kindly given by Douglas Wilson's daughter Mrs Hester Davenport, and the NASA Earth–Moon photograph (Plate 1) is reproduced by kind permission of Springer Scientific and Business Media from Head, J. W (2001) *Earth, Moon and Planets*, Vol. 85–86, published by Kluwer Academic Publishers. My task has also been made easier by the Cambridge University Press editorial staff, Martin Griffiths and colleagues, and their remarkably quick-responding referees who offered great encouragement and constructive help with the book's preparation at an early stage. To all these people I offer my sincere gratitude, at the same time accepting personal responsibility for any errors that may be apparent.

E. N.

1

Moonshine

Beliefs ... in an influence of the moon on life on Earth ... are by no means all moonshine.

H. Munro Fox, 1928

In 1957, Lamont C. Cole, a distinguished American ecologist, published an article entitled *Biological clock in the unicorn* (Cole, 1957). Surprisingly it appeared, not in a publication such as *The Annals of Improbable Research*, but in the prestigious scientific journal *Science*. It therefore had withstood critical scrutiny by peer review and was clearly of serious intent. In that journal, and with such a title, the article was guaranteed to make an impact, a precursor perhaps of scientific publications that in more recent years would compete for the Ig Nobel Prize for 'science that makes you laugh and then makes you think'. The paper was published at a time when a group of his fellow American biologists headed by Frank A. Brown, Jr were describing daily rhythms of biological activity in a range of living things, from intact animals to slices of vegetables. From the running activity of rats on treadmills, to colour changes of fiddler crabs and oxygen uptake by slices of fresh potato, from locomotor activity rhythms in marine molluscs to oxygen consumption by marine algae, hourly values were shown to increase and decrease throughout the day in regular rhythmic patterns that persisted even in so-called constant conditions in the laboratory (Brown, 1954, 1958; Brown *et al.*, 1955a, b). On the basis of such findings, a heated scientific debate was taking place as to whether the rhythms were controlled by internal physiological processes (biological clocks) or by subtle environmental variables, such as solar and lunar day changes in the earth's magnetic field and atmospheric pressure, and even daily fluctuations in the impacts of cosmic rays from space, that the experimental organisms in the laboratory were not shielded from. Brown and his co-workers favoured the view that daily changes in these residual geophysical variables could serve as time cues that triggered cyclical patterns of behaviour in living things and Cole's paper sought to present an unsubtle challenge to that view.

Brown (1965) presented a strong rebuttal of the Cole paper, referring to it as an 'unfortunate, very misleading publication'. He stated that the first studies of his group were based on the classical assumption that rhythmic periodicity was timed independently within each organism, as in Brown (1954) and Brown *et al.*, (1955a, b). For example, they reported persistent, seemingly internally generated, that is endogenous, daily and tidal rhythms of oxygen consumption in fiddler crabs. However, they quickly reasoned that if such rhythms were truly endogenous their patterns should vary significantly with temperature, which was found not to be the case (Brown *et al.*, 1954). This seemed most unexpected if biological timing was achieved by an internal physiological process so, one year after publication of the Cole paper, Brown (1958) again argued against the existence of internal biological clockwork. He claimed that he had 'incontrovertible evidence that even when we have thought we have excluded all forces influencing living things, there is, nonetheless, cyclic information, unquestionably with all the natural periodicities of the atmosphere imbedded in it, still impressing itself upon the organism'. Later, Brown (1962a, b) still considered that there was no evidence available to suggest that organisms possess independent timing mechanisms, but that they are 'dependent for their timing upon continuing response to the subtle rhythmic geophysical environment', referring to 'extrinsic rhythmicality' as a 'reference frame for biological rhythms under so-called constant conditions'. Brown and his co-workers therefore challenged the views of many other biologists at that time who were coming to accept the idea of endogenous clock-controlled biological rhythms. Brown's proposal was provocative, but the notion that animals, and even plants, possess some form of internal time sense had been in favour for much longer.

The French astronomer Jean Jacques de Mairan, early in the eighteenth century, incisively noted that the daily opening and closing of the leaves of *Mimosa* continued for several days in a nearly normal manner, even when plants were kept away from daylight cues in a continuously darkened cellar (see Daan, 1982). Later in that century Henri-Louis Duhamel du Monceau successfully repeated the experiment, keeping temperature constant, too, by observing his plants in a wine cave (see Winfree, 1987). His, and other, early interpretations as to how such rhythms were controlled considered that plants and animals learned, during their lifetime, to perform rhythmic behaviour that matched cyclical changes in their environment. Organisms were thought to establish daily patterns of behaviour in response to the benefits obtained by reacting in a particular way at one stage of the day/night cycle but not at another. That interpretation was strongly supported early in the twentieth century and, by the mid-twentieth century, it was extrapolated by some authors to suggest that time-keeping ability could be an inherited characteristic of living organisms.

Interestingly, this now fairly generally accepted view is contrary to that of Charles Darwin (Darwin and Darwin, 1880) himself, who had earlier questioned how rhythmic behaviour could have arisen by natural selection. He was particularly concerned as to how to explain the daily pattern of leaf opening and closing behaviour in plants. In fact, it remains the case for many animals and plants, that the selective advantage of innate rhythmicity has yet to be demonstrated experimentally. However obvious it may appear to be, it is difficult to prove that organisms with supposed inbuilt clockwork have a greater potential for survival over evolutionary time than individuals of the same species without such a timing mechanism, a point which will be discussed later.

It was perhaps because of uncertainties concerning how natural selection could favour inherited biological clocks that Brown (1958, 1960, 1962a, b, 1965) refocussed the debate, arguing that proponents of the internal biological clock hypothesis had not carried out sufficiently rigorous experiments. Even in 'constant conditions' in the laboratory, which usually meant controlled light, temperature, humidity and, for coastal animals, an absence of the influence of tides, living things, he argued, would still be exposed to some environmental variables that were not kept constant. He therefore concluded that organisms in such conditions were in fact responding to residual geophysical variables and were not behaving in response to their genetic makeup. Brown's interpretation persisted in some quarters, despite crucial evidence to the contrary that had been obtained even as early as the nineteenth century. Indeed, some of his more provocative findings were still being quoted as fact on the Internet in the early twenty-first century, despite serious questioning of their validity in the scientific literature after they were first published (see Chapter 5).

In the early 1800s, following up the earlier experiments by de Mairan and du Monceau, further studies of the leaf-opening rhythm of *Mimosa* were carried out in Geneva by Augustin Pyramus de Candolle. He found that in continuous dim light the plants exhibited not a 24-h rhythm of leaf opening and closure, but one of 22 to 22.5-h periodicity (see Winfree, 1987). Then, a century later, in the 1920s, similar evidence emerged concerning the behaviour of another plant, *Canavalia*. The periodicities of 'daily' rhythms of leaf movements of this plant in constant light, like those of *Mimosa*, were shown by Anthonia Kleinhoonte in Holland to deviate slightly from 24 h, at periods quite different from any known subtle geophysical variable that might be considered as a candidate to drive such rhythms in normal constant environment rooms in the laboratory (Kleinhoonte, 1928). Evidence such as this fostered scepticism of Brown's interpretation in the 1950s to the extent that Cole (1957) was prompted to test directly the hypothesis that biological rhythms might occur, not only in the absence of the more obvious environmental variables, but also when an animal was deprived of exposure to subtle geophysical variables too. Accordingly he elected to search for biological rhythms in an animal that could not possibly be influenced by such earthly factors, the mythical unicorn being ideal for his purpose.

Since Brown and his co-workers had demonstrated patterns of daily variation in the metabolic rate of, for example, potatoes, seaweed, carrots, earthworms and newts, the first task for Cole was to acquire data on the metabolism of his mythical animal. To do this he took a sequence of 120 values from a table of random numbers and postulated that they were consecutive hourly values of the standard metabolic rate of a unicorn kept for five days completely isolated from cycles of environmental variability, as of course they usually are. All that was then required for Cole was to use Brown's methods of statistical analysis of the time-series of its 'oxygen consumption' data to show whether or not the unicorn exhibited repeated daily changes in metabolism during its five days of sensory deprivation. Amazingly, from this analysis, the mythical unicorn did appear to show a consistent daily pattern of metabolic activity; its metabolism was greatest at night and least in the afternoon!

The point of the story lies in the fact that Cole used Brown's method of analysis of sequences of hourly values of biological data. Time series analysis of sequential data in the twenty-first century is now more sophisticated and objective but, in the 1950s, some procedures evidently succeeded in generating apparently meaningful biological rhythms, even in random data (see Enright, 1965a). The problem arose because Brown used data manipulation procedures that involved packaging sequential values of oxygen consumption into various sub-sets that could then be pooled for mathematical analysis. In his search for biological responses to residual geophysical variables, the challenge for Brown was to distinguish the responses of an organism to two daily patterns of change in these small scale changes that are of similar periodicity. These are of 24-h (solar day) periodicity, associated with the earth's rotation in relation to the sun,

and of 24.8-h (lunar day) periodicity associated with the perceived rotation of the earth in relation to the moon. Packaging of hourly data values
to search for lunar day rhythms involved 'data-slipping', which was considered by Cole to be a major factor in generating apparent rhythmicity
in random data. In his publication Cole (1957) was clearly questioning the
validity of the concept of biorhythms linked to residual periodic variables,
and also issuing a challenge to chronobiologists to use more rigorous statistical methods when seeking to demonstrate rhythmic phenomena in
plants and animals.

Whilst it is known, for example, that some animals appear to orientate to the pattern of the earth's magnetic field when navigating over
long distances (see Chapter 5), no direct evidence has yet emerged to
support the view that cyclical changes in the energy fields of residual
geophysical variables are detected and used solely by organisms as external (exogenous) daily time cues such as to preclude recognition of the
existence of internal biological clocks. Nevertheless, the old controversy
about whether biological rhythms are driven by endogenous or exogenous factors was later re-opened by Martin and Martin (1987). These
authors raised the possibility that the 24-h behavioural rhythmicity of
honeybees is not controlled by an endogenous, that is circadian, clock,
but is controlled solely by exogenous factors. Specifically, it was proposed
that bees respond to diel changes in the geomagnetic field and do not
possess internal biological clocks. That interpretation was challenged
immediately by Brady (1987), starting from the premise that most workers in the field of chronobiology must have thought the debate long since
settled. Brady re-examined the data of Martin and Martin concluding
that they were consistent with the more generally accepted explanation
of endogenous circadian rhythmicity, though acknowledging the proba-
bility that bees do use local magnetic cues to set their feeding rhythms
(see also Chapter 10). Certainly, in recent decades, a plethora of examples among plants and animals has emerged to confirm the endogenous
nature of many rhythmic biological phenomena, which free-run at periodicities that only approximate to environmental cycles when deprived
of such cues in the laboratory. In fact, since the mid-twentieth century,
research on the nature of biological rhythms has proceeded by seeking
answers to the following questions:

1. Do organisms in general show behavioural and physiological
    rhythms when maintained in constant conditions?
2. What are the properties of such rhythms and how persistent are
    they?

3.   What and where is the physiological clock which controls the approximate periodicity of the rhythms being investigated?

4.   What is the role of environmental variables and how, in nature, do they effect accurate phasing of endogenous physiological clocks?

5.   What is the ecological significance and, hence, the adaptive value of rhythmic behaviour and physiology in animals and plants?

* * * *

The concept of the 24-h biological clock is now in common usage, as recognition of the phenomenon of jet-lag in air travel confirms. Humans carry a body clock as they move around the earth and sense that resetting the clock in a new locality takes a longer or shorter time depending on the longitudinal distance travelled. Indeed, it can be said that the concept of the human body clock can be confirmed by experiment since by ingesting medicinal melatonin the clock resetting process can be speeded up. A physiological basis for such clockwork has been demonstrated in many animals including humans, clockwork which when deprived of external time cues is expressed as circadian rhythms of sleep/waking and various bodily functions. The word 'circadian' is also in general acceptance and quite common usage now; it defines rhythms of periodicities that approximate to the 24-h day and which are expressed in the absence of the external, clock-setting environmental cycles. The nature of circadian rhythms is best understood if the word circadian is pronounced *circadian* (literally: approximately daily) and not *cir-cay-dian*, which is most commonly the case.

It is, then, easy to believe that many living organisms on the spinning earth possess their own internal biological clocks (see also Dunlap *et al.*, 2004). General acceptance is such that the subject once attracted over-optimistic interest from stock market speculators and others seeking to prosper from the new science of 'Chronobiology'. In the early days of biorhythm conferences, speculative funding support came from financiers anxious to find out if economic cycles and trends in stocks and share values could be predicted by the new science. The only reasonable prediction was that the funding source would dry up quite quickly when it became apparent that chronobiologists had little to offer stock market speculators in exchange for conference-funding support. Other speculators, however, did succeed in marketing charts, books and even pocket calculating machines extolling what came to be known as the 'Biorhythm Theory'. The theory was based on suppositions that found their way into the medical literature that human physical ability waxes and wanes every 23 days, emotional condition varies on a 28-day cycle and intellectual performance follows a weak 33-day cycle. It was taken sufficiently seriously in the 1970s to be put to the test by airlines, the military and the insurance industry, all anxious to know if there was anything in it for them. Not surprisingly, perhaps, in all cases the theory was found to be wanting, leading Winfree (1987) to point out, that 'Biorhythm Theory' is no more than pseudoscience, with no justification whatsoever. Biorhythm Theory is to chronobiology what astrology is to astronomy, and it has done nothing to advance the credibility of the science of biorhythmicity. By contrast, chronobiology impacts significantly on medical research into sleep and other human rhythms, and on medical chronotherapeutics, which is concerned with the optimization of times of application of drugs (Winfree, 1987).

So, the strength and repeatability of daily changes in solar illumination can reasonably be assumed to have provided, over geological timescales, a suitable environmental backdrop against which circadian clocks have evolved, not only in humans, but also in a wide range of animals and plants. Indeed, the subject has wide implications for the exploitation and management of the world's natural resources. Sustainable exploitation in agriculture and fisheries, optimization of conditions for artificial cultivation of living organisms, and the development of rational strategies for environmental conservation, should all be founded on a full understanding of the cyclical nature of living animals, plants and ecosystems (Naylor, 2005). Is it the case then that the repeatability of lunar cycles, like that of solar cycles, has also provided a consistent environmental background against which tidal time-keeping ability has been selected for during the timescale of biological evolution? Certainly, many biologists, before and after Cole, have expressed reservations about moon-related activities in animals. However, direct responses to moonlight, particularly in relation to the lunar monthly cycle, but possibly also in relation to the lunar day, may not be so far-fetched as is sometimes assumed, as will be discussed in Chapter 5. Moreover, indirect effects of the moon reflected in the tidally related behaviour patterns of marine animals are now well understood and it is the nature of ocean tides which provide the evolutionary backdrop to those behaviour patterns that will now be considered.

*   *   *   *

Ocean tides on earth are caused primarily by the gravitational pull of the moon, and to a somewhat lesser extent by the gravitational effect of the sun (see Pugh, 1987) and they vary according to the moon's orbit around the earth and the combined earth–moon orbit around the sun (Figure 1.1; Plate 1). The moon's gravitational force draws up a bulge in the sea lying beneath it, which is balanced by a reciprocal, centrifugally

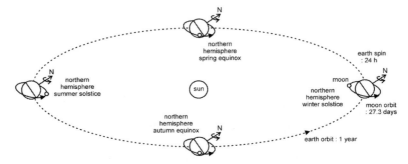

Figure 1.1 Earth tilt and spin, tilted orbit of the moon, and earth–moon orbit around the sun.

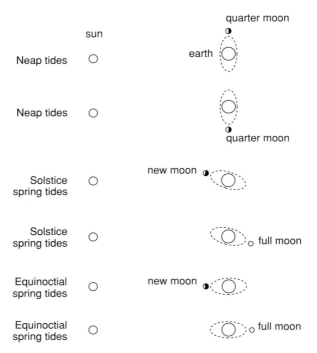

Figure 1.2 Snapshots of relative positions of the moon and sun that generate neap tides, solstice spring tides and equinoctial spring tides, showing exaggerated tidal bulge (dotted) in each case.

generated bulge in the sea on the opposite side of the earth (Figure 1.2). The centrifugal force that generates the reciprocal bulge arises because the earth and the moon are effectively spinning as one mass, the common centre of gravity of which is not the centre of the earth but a point nearer the earth's surface in the direction of the moon. Since the distortion of the ocean remains stationary in relation to the moon, the daily

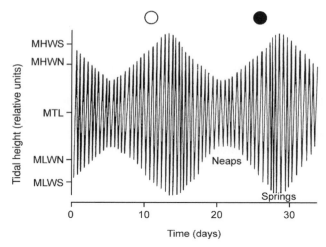

Figure 1.3 Generalized pattern of semidiurnal tides (12.4 h intervals) over a lunar month. MHWS, mean high water spring tide level; MHWN, mean high water neap tide level; MTL, mid tide level; MLWN, mean low water neap tide level; MLWS, mean low water spring tide level; open circle, full moon; closed circle, new moon (after Naylor, 1982).

rotation of the earth ensures that an upward bulge in the sea, that is a high tide, appears at any one point on the earth's surface twice every day. If the moon was stationary above the earth the interval between successive bulging high tides would be precisely 12 h, as a function of the 24-h rotation of the earth. But, of course, the moon itself is also moving, on its monthly orbit around the earth. This displacement, together with the daily rotation of the earth gives to an observer at any point on the earth an apparent orbit of the moon around the earth of 24.8 h. This defines the lunar day, as distinct from the 24 h solar day. So, with a lunar day of 24.8 h, and two ocean bulges typically passing during that time, the condition of high tide is generated at any one point on earth every 12.4 h, that is twice every lunar day, defined as semidiurnal tides (Figure 1.3). Such tides would be readily predictable from the position of the moon overhead if the earth was completely covered by an ocean of uniform depth, if the orbit of the moon was exactly in the plane of the earth's equator, if the plane of the earth's equator was exactly in the plane of the earth's orbit around the sun, and if other things were equal; but none of these requirements prevails.

First it is important to note that the orbit of the moon is tilted in relation to the earth's equator. Twice during each of its monthly (27.3 day) orbits around the earth the moon passes over the equator but, between times, the declination of its orbit is such that it is directly overhead in the northern hemisphere during part of its orbit and, about 14 days later, is overhead for several days in the southern hemisphere. As a result, the tidal bulge beneath the moon, and its reciprocal on the other side of the earth, oscillate northwards and southwards of the equator throughout the lunar month. This gigantic oscillation means that in some latitudes and at certain times of a lunar month, a particular locality on the spinning earth will be exposed to the deep part of one tidal bulge and to a shallow part of the opposite bulge 12.4 h later. Consequently, such localities experience more-or-less equal semidiurnal tides at those times of the month when the tidal bulges align with the equator and, at other times, alternate tides will be somewhat different in height. Geographical regions of the earth that experience such alternating patterns of tidal oscillations are said to experience mixed semidiurnal tides. Indeed, in some localities the inequalities may be so extreme as to produce only one high tide every 24.8 h.

Other factors, too, further complicate the tidal picture. Continental and island land masses interrupt water flows as the tidal bulges pass, and lateral flows may be exaggerated depending upon inshore depth and coastline shape, particularly along straits and estuaries. In addition, Coriolis force, generated by the rotation of the earth, deflects water movement to the right in the northern hemisphere and to the left in the southern hemisphere, generating local differences in the timing of high tide.

Also, strikingly in many localities, there are lunar monthly variations in the height of high tides. These are well known to fishermen and small boat owners who, if forgetful, may for several days on two occasions each month, find their craft high and dry, that is neaped, above high tide. To understand these monthly changes in tidal height it is necessary to consider not just the interaction of the earth and moon on a daily basis, but also their relationship to each other and to the sun over the lunar month. It is also necessary to take into account the additional gravitational effect of the sun on ocean tides and, therefore, the combined effects of lunar and solar gravity. Though the sun is vastly larger than the moon, it is so far away that its gravitational effect on earth ocean tides is only about half (46%) of that of the moon, but it nevertheless has a significant impact.

The time taken for the moon to orbit the earth, 27.3 days, is defined as one sidereal month. After that time, the moon returns to the same point overhead relative to the earth. However, because the earth and the moon together have also moved during this time in their orbit around the sun, the moon does not regain its position on the line between the earth and the sun for another 2.2 days. Consequently the interval between successive full moons as viewed from the earth is 29.5 days, defining the true lunar or synodic month. It is this cycle of events that normally determines monthly variations in tidal height, but see Chapter 5 for an exception. At full moon, when the moon is on the opposite side of the earth from the sun, indicated by the fact that it rises in the east as the sun sets in the west, the combined gravitational pulls of the sun and moon enhance the standing wave of the ocean bulges. Similar enhancement of the ocean bulges occurs at the times of new moon when the moon is almost directly in line between the earth and the sun and we see little of it except for the small crescent of sunlight that it succeeds in reflecting. Thus for about five days around times just after the full and new moon, at intervals of 14–15 days, high (and low) tides are most extensive, defined as spring tides (Figures 1.1, 1.2, 1.3) from the Old English word 'springan' meaning to leap up. In contrast, in the intervening periods, when the sun and moon are at a right angles in relation to the earth and the moon is in its first and third quarters, the gravitational pulls of the sun and moon oppose each other. At these times, halfway between the times of spring tides, the tidal bulges are at their lowest, producing lower tidal ranges than at springs, so-called neap tides, from the Old English 'nepflod'.

The phenomenon of close alignment of the sun, moon and earth that occurs twice each synodic month at the times of full and new moon is known as syzygy. The effects of syzygy might be thought to produce spring tides of almost equal heights at new and full moons, and this would be so if the moon's orbit around the earth was circular and in the plane of the earth's equator. However, the orbit of the moon is not only tilted as we have just seen, but is slightly elliptical too. Both of these irregularities determine that the distance between the moon and the earth varies slightly during the lunar orbit, resulting in varying gravitational effects on ocean tides. Most noticeably, they generate differences in height between a set of new moon tides and the following full moon spring tides. New moon tides are sometimes higher and sometimes lower than the succeeding full moon tides, the largest spring tides alternating between new moon and full moon spring tides approximately every seven months. This switching of the timing of maximal spring tides between full and new moon phases arises because of the interplay between the synodic (syzygic) cycle of lunar events and the cycle of lunar proximity to the earth. Occasions of closest proximity of the moon to the earth (perigee) occur at intervals of 27.5 days. This interval, known as the anomalistic month, approximates in duration to the sidereal month but is two days shorter than the synodic month with which it interacts. Perigee and syzygy coincide approximately every seven months, and it is at those times that the switching occurs.

Finally, even more variation in the range of spring tides is apparent over the year; they are most extreme at the spring and autumnal equinoxes when the earth, moon and sun are most perfectly in line at the times of new and full moon. This arises because of the tilt of the earth's axis of spin which points the earth's northern hemisphere towards the sun during the boreal summer and away from the sun in winter, during the austral summer. During these times of solstice the orbit of the moon around the earth is similarly tilted in relation to the sun, so that at such times the moon never quite crosses a line drawn from the sun to the earth and beyond. However, at the equinoxes the moon's orbit transits the earth's equator exactly at those times of the year when the sun is overhead at zero latitude. This close alignment of the sun, moon and earth generates equinoctial tides (Figure 1.2), which, if they coincide with storm surges, present serious flooding threats to coastal towns and cities.

All the factors discussed here ensure that there is considerable variation in the pattern of ocean tides depending upon the precise locality where they are experienced. Thus, whilst the average interval between times of high tide is usually 12.4 h, tidal range varies according to the phases of the moon and to locality. Also, the time of high tide may vary over quite small geographical distances depending upon coastline topography and water depth. Around British coasts and in many other places around the world, tides are typically semidiurnal, with successive high tides of similar ranges that increase at springs and decrease at neaps. The passage of the tidal bulge around the British Isles determines that the times of high tide vary over relatively short distances such that, for example, high tides along the coasts of south Wales consistently occur about 6 h earlier than along the north Wales coast only 300 km distant around the coastline. Extreme ranges of tide of more than 10 m are seen in the narrowing estuary of the Bristol Channel, UK, and in the Bay of Fundy, Canada, whilst tides are negligible in enclosed seas such as the Baltic and Mediterranean.

Mixed semidiurnal tides occur in a number of localities around the world such as the south west coast of the USA, the Philippines and Hong Kong. In these areas tidal patterns vary regularly throughout the lunar cycle, from two distinct 12.4 h tides each day on some days, to intervals when the two high tides appear to merge. The merger may be partial or complete, in which case, for a few days each month, the tidal intervals are 24.8 h. On the San Francisco coast of south west USA the 12.4 h form of tide dominates the mixed semidiurnal pattern, but around Manila in the Philippines the 24.8 h form of tide predominates. Off the coast of Vietnam, the tides are of the full diurnal type; that is they are of 24.8 h periodicity, again varying in range over the lunar cycle from springs to neaps.

Notwithstanding the fact that some localities exhibit tides of 24.8 h periodicity, the most common expression of ocean tides around the world occurs to a greater or lesser extent at 12.4 h intervals, two high tides occurring during each lunar day. It is the case that in some localities, for geophysical reasons, the timing of a particular high tide may be an hour or so longer than 12.4 h, but that is followed by a tide of an equivalent length of time shorter than 12.4 h, successive tides alternating their periodicity slightly around the 12.4 h standard. Similarly, variations in atmospheric barometric pressure, and wind speed and direction, may induce slight changes in the precise timing of high tides, but again, the integrated high tide signals to animals and plants living on most of the world's tidal coastlines are of 12.4 h periodicity, occurring twice during each lunar day. Indeed, the repeatability of tidal signals for measurement of the passage of time appears also to have been used by humans since long before understanding of the causes of tides became known. As Pugh (1987) points out, the word 'tide' derives from the Saxon word 'tid', meaning 'time', as also used in defining seasonal festivities such as Eastertide and Yuletide.

*   *   *   *

Since we are considering the possibility of evolutionary adaptation by living organisms to the timing of tides, it remains to ask if and how tidal intervals have varied over geological time. To address these questions, it is necessary to consider changes in the rate of the earth's rotation and the rate of increase of the radius of the lunar orbit around the earth. Recent estimates suggest that the daily rotation of the earth is currently slowing down by about 2.5 ms every hundred years. Even over intervals of thousands of years this factor is probably trivial in the context of biological variability, but it has relevance over the hundreds of million years of evolution of living organisms, since it has resulted in a gradual lengthening of the solar day. Moreover, it has also been estimated that the moon is retreating from the earth at the rate of about 5.8 cm every year, thus lengthening the circumference of its orbit. Again this has a trivial effect on the lunar cycle over intervals of thousands of years, but might be expected to have some effect on tides over geological time. Indeed, on the basis of factors such as these it has been estimated that during the late Palaeozoic Era, some 300 million years ago, the earth was spinning faster than at present and the moon was orbiting faster such that the length of a solar day was about 20 h and the duration of the lunar month about 34 of such days, equivalent to 28.7 days as measured in present 24 h day lengths. These estimates suggest that there would have been more than 400 Palaeozoic days in a year-long orbit of the earth around the sun at that time. Indeed, fossil corals from the Palaeozoic provide independent evidence of those estimates; close analysis of their growth rings shows that they possess around 400 presumed daily rings within each annual ring (Jones, 1999). In the Palaeozoic Era then, the lunar cycle was a little shorter than the present 29.5-day lunar month, and this shortening, coupled with a short solar day of 20 h, would have generated tidal bulges that progressed, not at intervals of 12.4 h, but of more like 10.3 h. In addition, since the moon was closer to the earth then, than it is at the present time, the tides would also have been larger than they currently are, perhaps by as much as 10–20%, depending upon local coastal topography.

* * * *

There is no doubt that the pattern of tides is somewhat variable over the global ocean, and may vary quite substantially from day to day in some individual localities. On this basis, it is sometimes suggested that the variability is such as to make it unlikely, or at least surprising, that coastal animals could have evolved biological clockwork permitting them to anticipate tidal rise and fall. Comparisons are sometimes made with the occurrence of circadian rhythms in terrestrial animals that can more understandably be seen to have evolved against the seemingly more predictable backdrop of the 24-h cycle of light and dark. It is suggested that, having evolved biological clocks that approximate in periodicity to the day/night cycle, those clocks could also be used in some way as adaptations to the tidal cycles in the environment. After all, two tidal cycles themselves also approximate to the 24-h cycle of day and night. There could be some cogency to this suggestion in localities in the world where mixed semidiurnal tides occur, particularly in the few localities where a single high tide occurs daily at some times of the month.

However, even in such localities the fundamental periodicity of tides is 12.4 h, as it is in many other localities around the world where tides are typically semidiurnal, and have been so over recent millennia. The tidal interval has increased since life on earth began, but so also has the length of the solar day, and the rate of increase of each has been very slow. Animals, as distinct from single-celled organisms, first colonized the marine intertidal zone over 500 million years ago. The lengthening of the tidal interval has been, on average, about 1 s every 40 000 years and there is no reason to suppose that genetic adaptation could not keep pace with lengthening tidal and daily cycles over geological time.

Thus, notwithstanding sporadic and minor variations in the timing of tides, attributable to wind and barometric pressure, one can say that coastal marine organisms living on tidal shores are exposed repeatedly to the day/night cycle, and also to the complex environmental changes induced by the rhythm of tides. It therefore seems reasonable to suppose that animals have adapted to life in tidal environments in such a way that they have evolved not only daily biological clocks, but that the strength of the tidal signals throughout evolutionary history has led to the acquisition of tidal clocks too. Evidence that coastal animals possess such clocks which are expressed as circatidal, that is approximately tidal rhythms, in the laboratory will be considered in the next chapter. A later chapter, too, will consider whether there is sufficient scientific justification to show that coastal animals are able to respond directly to the light of the moon. At this stage we have at least built up a background of the lunar influences on ocean tides, against which biological rhythmicity might reasonably be expected to have evolved.

2

Biorhythms of coastal organisms

*...whilst this planet has gone cycling on according to the fixed laws of gravity
...endless forms most beautiful and most wonderful have been, and are being,
evolved.*

Charles Darwin, 1859

Throughout Europe in the mid-nineteenth century there was intense scientific curiosity about the communities of animals and plants that were revealed by the rise and fall of tides along Atlantic coastlines. In Britain, Charles Darwin, Thomas Huxley and Philip Gosse were leading Victorian naturalists inspired to study that fauna and flora. Later in the nineteenth century, not content to merely observe, describe and classify living organisms in coastal seas, naturalists also saw the need to try to understand how they functioned and behaved. To achieve such objectives it was soon apparent that they would require facilities in which to maintain marine animals and plants under near-natural conditions in seawater aquarium systems. Cogent advocacy of their requirements, at a time of burgeoning public interest in marine life, then led to the establishment of specialist laboratories throughout the world in which detailed observations and experiments could be undertaken on a wide range of marine animals and plants (Ryland, 2000).

One such laboratory established by French biologists in 1872 was at Roscoff, in western France. There, on the beaches of Brittany, French and visiting biologists first recorded a striking rhythmic phenomenon, subsequent study of which proved to be seminal in the development of our understanding of biological clocks in marine animals and plants. When walking along Roscoff beaches at low tide, early naturalists observed that patches of green colouration appeared on the sand surface as the tide ebbed, only to disappear again before the tide returned to cover them. Some of the earliest, scientifically documented, accounts of this phenomenon were published in the prestigious *Proceedings of the Royal Society of London* (Gamble and Keeble, 1903). These authors showed that the green patches on the sand surface at low tide were caused by the emergence of large numbers of small, burrowing flatworms, appropriately named *Convoluta roscoffensis*. The green colour was found to be due to the presence in the flatworms of zooxanthellae, single-celled green plants that live symbiotically in the tissues of the worms. Emergence of the flatworms on to the sand surface during daytime low tides was clearly of benefit to the zooxanthellae by exposing them to the daylight they required to utilize their green chlorophyll in the process of photosynthesis. Also, in true symbiotic fashion, the flatworms benefited from the presence of the zooxanthellae by absorbing their waste products as nutrients. However, a particularly striking feature of this primitive animal/plant association was the behaviour of the flatworms in re-burrowing into the sand, apparently in anticipation of the rising tide. Their burrowing behaviour ensured that they were not washed away from their optimal zone on the beach, just above mid-tide level, and obvious questions were quickly apparent as to how their tidal timing was achieved. Did these simple flatworms burrow in response to the vibrations of breaking waves during the rising tide, or had they some internal sense of time? And, if they possessed a sense of time, was it in the form of a countdown timer that measured time from the moment that they emerged after exposure on the falling tide, or was it a more sophisticated clock system even than that? More to the point, was there a way of finding answers to these intriguing questions? In the event, the experimental procedures adopted to find the answers were essentially those still initially used to solve similar marine biological clock problems today.

It so happened that in the same year that Gamble and Keeble published their observations on the behaviour of *Convoluta*, the results of laboratory experiments on the remarkable behaviour of this flatworm were published by Bohn (1903). Seeking an answer to the question as to whether the flatworms burrowed in response to the physical conditions generated by waves and tides, Bohn carried out the simple experiment of isolating freshly collected worms in sand in laboratory aquaria. He showed that the flatworms continued with their pattern of emergence onto the surface of the sand at the approximate times of low tides, re-burrowing again before the times of high tide, even when denied the influence of tides. The rhythm continued in a continuously lit aquarium room, persisting for about eight 'tidal' cycles, that is for four days in constant conditions. After that time the flatworms continued to emerge at intervals onto the sand surface in the aquaria, but in a random, arrhythmic manner. Critically, Bohn then carried out the reciprocal experiment, returning flatworms to the beach in a container from which they could not escape yet were again exposed to the rise and fall of tides. Then, after a few days back in their home environment, he took the flatworms into the laboratory again to find that their apparently spontaneous tidal rhythm of behaviour had returned. On the basis of this work, Bohn suggested that the flatworms were perhaps in some way able to measure time and not simply as though they were using an hourglass or countdown type of timing mechanism. Still later, with a colleague, he published a further paper expanding on his earlier view referring to the rhythmic surfacing of *Convoluta* as 'la phenomene de l'anticipation reflexe' (Bohn and Pieron, 1906). In these two papers, probably for the first time, Bohn provided experimental evidence for the existence of true biological clocks of approximately tidal periodicity in marine animals. Then, a year later, Martin (1907) successfully repeated Bohn's experiments and confirmed those findings in a paper entitled 'La memoire chez *Convoluta roscoffensis*', strongly supporting the notion that this diminutive intertidal flatworm possessed a tidal memory.

Surprisingly, Keeble (1910) who, with Gamble, had first characterized the tidal behaviour of *Convoluta* from his observations of their migrations on the beaches at Roscoff, was not able to accept the idea that the behaviour pattern might be attributable to an anticipation reflex, with its implied measurement of time. In challenging this concept and by advancing the view that time measurement was achieved by reference to external signals, he initiated the debate referred to in Chapter 1. His views were subsequently supported by Frank A. Brown and colleagues working at Northwestern University and at the Marine Biological Laboratory, Wood's Hole, in the USA (Brown, 1962a, b, 1965), who continued to champion the notion that biological rhythms were primarily driven by exogenous factors. In fact, as Morgan (2001) pointed out, the early descriptions of the behaviour of *Convoluta* formed the cornerstones of the argument in favour of the importance of endogenous, that is, internal control of biological rhythms, as presented by Pittendrigh (1960).

Clearly, from the findings of Bohn and Martin, *Convoluta* was not able to keep its clock running indefinitely in the laboratory, which it might be expected to do if it relied on external time signals. Its tidally rhythmic behaviour persisted for only a few days in constant conditions, after which the pattern of behaviour became random. Moreover, as will be apparent later, there are now many more examples of such tidal

memories in coastal animals kept in constant conditions in the laboratory that are expressed in exactly the same manner as that of *Convoluta*.
In all cases, the general rule is that the spontaneous rhythmic behaviour
shown by such animals only approximates to the periodicity of tides,
that is, it is circatidal and, like circadian rhythmicity, it changes to a random pattern after some days. Both circatidal and circadian rhythms are
expressed at frequencies quite different from any known geophysical variable, suggesting that it is unlikely that they are driven by environmental
variables when expressed in conventional controlled environment rooms
in the laboratory.

A second important strand of the argument against the concept
of tidal memories and in favour of exogenous control of the rhythmic
behaviour of animals, concerned the fact that, in some instances, the
timing of biological clocks appeared to be independent of the temperature to which animals were exposed in the laboratory. Whilst this might
not be surprising in warm-blooded animals that control their body temperature within a limited range of environmental temperatures, it was
surprising that it occurred even in cold-blooded animals such as fiddler
crabs (Brown *et al.*, 1953), whose biorhythms had been studied extensively by Brown and his co-workers. Within limits, the body temperature
of a cold-blooded animal adjusts to the temperature of the surrounding environment, the rate of the animal's metabolism speeding up or
slowing down accordingly. So, if such an animal possessed an internal
physiological clock one might expect the clock to run faster at higher
temperatures and slow down if the temperature dropped, resulting in
expressed biorhythms of shorter or longer wavelengths, respectively.
However, this did not happen when Brown and his colleagues recorded
the locomotor rhythms of fiddler crabs at different temperatures in the
laboratory (Brown *et al.*, 1953; Brown, 1962a, b, 1965). They, and many
other biologists working with other cold-blooded animals, found that,
within the normal range of temperatures tolerated by the animals concerned, the wavelengths of the animals' biorhythms remained the same.
In other words, their body clocks, if they have them, remain unaltered
in their timing despite the speeding up or slowing down of their general metabolism with rising or falling temperatures. For Brown, notwithstanding the mismatch between periodicities of biorhythms in the laboratory and environmental geophysical cycles, the simplest hypothesis to
account for the temperature independence of biological rhythms was
that the rhythms were not controlled by internal physiological processes. It was argued that since biological clocks function normally in
plants and cold-blooded animals irrespective of temperature, both in the

environment and in the laboratory, then timing signals must come from outside, and not from an internal biological clock. But even this apparently very strong point in favour of exogenous control of biorhythms could be countered if it could be shown, for example, that cold-blooded animals such as crabs had temperature-compensating physiological clockwork, a possibility that will be considered in a later chapter.

Brown's interpretation of the nature of biological clocks in animals and plants provoked many counter arguments in the scientific literature. Not only Cole (1957) with his provocative paper on *Biological clock in the unicorn* but many other chronobiologists also vigorously rejected his ideas (Aschoff, 1960; Pittendrigh, 1960). The notion of true biological clocks, based upon the early studies of the migratory behaviour of *Convoluta roscoffensis*, steadfastly withstood the test of time. Nevertheless, in the 1950s, the debate still posed many questions and suggested interesting experiments that might be carried out. Moreover, marine coastal animals in localities with extensive twice-daily tides seemed to be particularly challenging experimental subjects for study considering the wide range of daily and tidal environmental variables to which they are exposed. One such widely distributed and hardy species in a British locality which seemed appropriate to work with at the time was the green shore crab *Carcinus maenas* (Naylor, 1958) (Plate 2).

At low tide, a hunt for this common crab begins by turning stones and boulders to search for crabs in their hiding places, where they usually remain inactive when the tide is out. It is well known, too, that on rising tides, around the British Isles, shore crabs are often seen on the open shore when their habitat is flooded, during which times they can be caught with baited lures and traps. The question arises, however, as to how the timing of their emergence at high tide and their return to shelter at low tide is achieved. Are they simply responding to immersion by the rising tide, and to emersion as the tide falls? Or do they have some form of time sense that awakens them during the flood tide and permits them to return to shelter as the tide falls? Or, indeed, is it the case that shore crabs normally remain hidden throughout much of the tidal cycle, waiting as ambush predators to emerge and capture food that comes to them in swirling tidal currents during high tide? A partial answer to such questions can be given by donning a face mask and swimming between tidemarks at various times throughout the tidal cycle. By this means, it can certainly be observed that large numbers of crabs move actively over the shore during rising tides, presumably in search of mates or to find food, and not behaving as ambush predators.

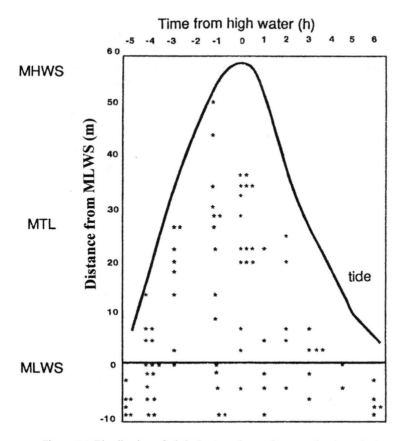

Figure 2.1 Distribution of adult *Carcinus* observed on a rocky shore during dive-transects throughout a spring tide cycle. Each asterisk represents a single crab observed within a 4 × 2 m rectangle of shore, plotted with respect to the distance from mean low water springs and the time in hours from high tide. MHWS, level of mean high water spring tides; MTL, mid tide level; MLWS, level of mean low water spring tides (after Warman *et al.*, 1993a).

More importantly, however, a team of divers observing and counting crabs throughout a complete tidal cycle recorded that *Carcinus* not only begins to range freely over the shore when the tide comes in but that it also seeks shelter under boulders before the tide leaves them high and dry (Warman *et al.*, 1993a). The results of such a survey are illustrated in Figure 2.1 which shows that the distribution of large crabs on the shore is skewed towards the time of the rising tide, few being seen near the tide edge during the falling tide. The crabs, like the flatworm *Convoluta*, therefore appear to anticipate low tide conditions, hiding when the tide is out, when otherwise they might be easy prey to their own predators such as gulls and other seabirds. Moreover, it is the case that free-roaming *Carcinus maenas* below tidemarks, monitored by radio-acoustic positional telemetry during warmer months of the year, are more active during high tides than low tides (Lynch and Rochette, 2007). All of these quantitative observations confirmed earlier anecdotal knowledge of the behaviour of crabs on the shore, on the basis of which it was possible to hypothesize, even in the 1950s, that crabs would persist with a tidal pattern of activity and rest even if they were isolated from tides in the laboratory.

To test that hypothesis, appropriate experiments were carried out by recording crab behaviour in the laboratory in conditions not dissimilar from those that they would experience for at least part of their time in the natural environment (Naylor, 1958). Importantly, too, it was necessary to pre-empt criticisms of similar experiments in this field that had been raised in the scientific literature. For example, at the time of the first experiments with *Carcinus*, related studies on other animals had been criticized because they were carried out with groups of individuals that probably influenced each other behaviourally, generating masses of potentially unreliable data that were analysed by questionable statistical procedures. Lamont Cole's article on *Biological clock in the unicorn*, which had so sharply criticized statistical procedures for time-series analysis, immediately raised awareness of the need to preclude such criticism as new experiments were designed. Initial experiments on *Carcinus maenas* were therefore carried out on individual crabs, each enclosed in a separate plastic box, and kept moist with a little seawater, simulating as near as practicable the conditions that a crab would experience under boulders at low tide. Each box, or actograph (Plate 3), was mounted as a see-saw within a separate frame, and a wire stylus beneath each box wrote the daily pattern of the crab's movements on a rotating drum beneath. A freshly collected crab was placed in each box at the start of an experiment, which was carried out in a continuously darkened room, at constant temperature and away from the influence of tides (Naylor, 1958).

In those initial experiments, individual crabs showed alternating episodes of walking activity and quiescence, as the initial hypothesis had predicted. Active sessions initially coincided with the times of high tide on the adjacent seashore and continued approximately in phase with high local tides for several days in the laboratory (Figures 2.2, 2.3). Evidently the

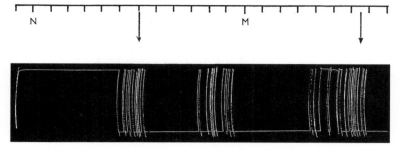

Figure 2.2 Actograph trace of a single *Carcinus maenas* collected at a time in summer when it had been experiencing both high tides during the hours of daylight and maintained in the laboratory at constant temperature, in continuous dim light and away from the influence of tides. Arrows, times of expected high tide; M, expected midnight (after Naylor, 1958).

urge to vacate their hiding places and forage during high tide persisted even in laboratory actographs, to the extent that, occasionally, a crab would prise open the lid of its box, normally held in place with rubber bands, and escape. Accordingly, some stylus records on the recording drums came to an abrupt end, always at the times of local high tide, necessitating the rescue of escaped crabs from the floor of the aquarium darkroom! The approximately tidal nature of their rhythmic activities did not coincide with any environmental variable that might conceivably have influenced the crabs in the controlled conditions of the laboratory. Therefore, as a preliminary confirmation of the hypothesis, each crab appeared to possess a tidal memory in the form of an internal circatidal clock mechanism that timed its pattern of behaviour in the laboratory.

In addition to their tidal bursts of walking activity, individual crabs in the laboratory were also more active around midnight (Figures 2.2, 10.8), even though they were provided with no cues as to the day/night cycle. This heightened activity was expressed as a burst of movement within the actographs around the time of midnight. At those times of the lunar monthly tidal cycle when high tides occurred around dusk and dawn the midnight burst of walking activity appeared as a third peak between the peaks at the times of 'expected' high tides (Figure 2.2). On the other hand, at times of the lunar monthly tidal cycle when expected high tides were around noon and midnight, the additional nocturnal waking session simply exaggerated the night-time high tide burst of walking activity of the crabs in the actographs. These nocturnal outbursts of activity were perhaps to be expected because, at low tide at night on a rocky shore, the crabs can sometimes be heard scuttling

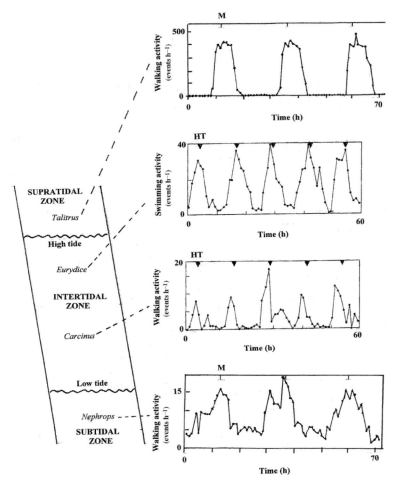

Figure 2.3 Spontaneous locomotor activity (events per hour) during 2–3 days in constant conditions in the laboratory of supratidal amphipods (*Talitrus saltator*), intertidal sand beach isopods (*Eurydice pulchra*) and shore crabs (*Carcinus maenas*), and subtidal Norway lobsters (*Nephrops norvegicus*). M, expected midnight; HT, expected high tide (redrawn from Bregazzi and Naylor, 1972; Naylor, 1985; Atkinson and Naylor, 1976).

noisily amongst boulders. By implication, too, as turns out to be the case, crabs forage more intensively on the shore during nocturnal high tides than during high tides that occur during daylight hours. The high tide and night-time patterns of behaviour of crabs in the laboratory clearly suggested that *Carcinus* possesses both circatidal and circadian biological clocks. The question then arises as to how generally such living clockwork

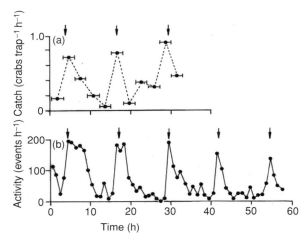

Figure 2.4 Field and endogenous locomotor activity rhythms of the crab *Helice crassa* on a muddy shore in New Zealand: (a) numbers of crabs collected in a grid of 20 traps (crabs trap^{-1} h^{-1}) emptied at 3 h intervals (horizontal bars) over three consecutive tides; (b) record of total hourly walking activity of five crabs in the laboratory in constant dim light at a constant 15°C during five periods of expected high tides (arrows) (after Williams *et al.*, 1985; by permission of Elsevier).

occurs among coastal organisms and how widely it is reflected in the physiology and behaviour of a single animal or plant.

* * * *

Life on earth probably dates back at least 3500 million years, during which time the earth has made some 1.5 million million turns on its axis (Daan, 1982). The number is huge, notwithstanding the slowing down of the earth's rotation by a few milliseconds every hundred years. Even so, however many rotations of the earth have occurred since living organisms first appeared on the planet, that very large number should be almost doubled for the tidal cycles that have influenced coastal regions throughout organic evolution. In coastal seas, where life itself probably began, daily changes in light intensity and temperature, and tidal changes in a wide range of environmental variables, have provided the physical background against which many strands of biological evolution have occurred since the origin of life itself. With such daily and tidal changes in environmental conditions between tidemarks it can reasonably be assumed that the processes of natural selection must have operated differently, not only between day and night, but also between high and low tide. If so, one can

expect that not only has 24 h rhythmicity become a pronounced characteristic of living organisms in general, but that tidal rhythmicity has also
become equally pronounced in organisms living in localities exposed to
the rise and fall of tides. Most obviously, it is the case that mobile organisms, in general, that live between tidemarks exhibit circatidal, and often
circadian, rhythms of locomotor activity when maintained under laboratory conditions (Palmer, 1974, 1995a; Naylor, 1985). Such rhythmicity has
been recorded for some years now in a wide range of marine organisms
from rocky and sandy shores, including, for example, diatoms (Figure
2.9) (Round and Palmer, 1966), flatworms (Gamble and Keeble, 1903),
polychaete worms (Last et al., 2009), amphipod and isopod crustaceans
(Figure 2.3) (Enright, 1963, 1965a; Jones and Naylor, 1970), crabs and
lobsters (Figure 2.3) (Naylor, 1958; Atkinson and Naylor, 1976), oysters
(Brown, 1954) and shore fishes (Gibson, 1965; Northcott et al., 1990). In
these examples and in many others, it is well known from observations or
from experiments that the circadian and/or circatidal rhythms expressed
in constant conditions in the laboratory more or less reflect patterns of
behaviour expressed by the organisms concerned in their natural habitat.
A typical example is that of the New Zealand mud crab *Helice crassa* (Plate
4) which exhibits a persistent circatidal rhythm of locomotor activity in
constant conditions, with peaks occurring at expected times of high tide,
the times when it is most active on its home beach (Figure 2.4) (Williams
et al., 1985). Rhythmic behaviour on the shore was demonstrated by catch-
ing mud crabs in pitfall traps set out on the shallow-sloping home beach
and emptied at 3-h intervals throughout three consecutive tides. Catches
were greatest during high tides, reflecting the foraging behaviour of the
crabs at those times. Indeed, endogenous rhythms of circadian and/or circatidal periodicity have long been considered to be universal phenomena
in eukaryotic organisms, from protozoa, fungi and algae to higher plants
and animals. Relatively recently, too, in the 1980s, circadian rhythmicity
was demonstrated in some cyanobacteria, formerly referred to as blue-
green algae. These unicellular organisms are prokaryotes, lacking the
membrane-bound organelles of eukaryotes, and were previously thought
not to possess circadian clockwork. Until the 1980s the 'membrane model'
(see Chapter 10) was considered to be an attractive explanation for circadian clockwork, which was thought to occur exclusively in eukaryotes.
Prokaryotes, with their lack of membrane-bound organelles and rapid
rates of cell division, were considered to be unlikely exponents of such
rhythmicity (Sweeney, 1976). Now it is apparent that circadian rhythms of
cell division are present in several prokaryotic cyanobacteria, including
a marine species of *Synechococcus*, which shows temperature-compensated

circadian rhythms of cell division (Sweeney and Borgese, 1989). In fact, as will be discussed in Chapter 10, such organisms, together with some unicellular eukaryotic algae, serve as model systems for the analysis of circadian rhythms at the cellular level. Two particular examples of unicellular marine algae which have been known for some time to exhibit temperature-compensated circadian rhythms are *Lingulodinium polyedrum* (formerly *Gonyaulax polyedra*) (Hastings and Sweeney, 1957; Sweeney and Hastings, 1957) and *Acetabularia* (Sweeney and Haxo, 1961).

Notwithstanding these early findings it remained for many animal species, and certainly in the case of the shore crab *Carcinus*, to be shown unequivocally that rhythms recorded in so-called constant conditions were not controlled by exogenous factors. After all, when experiments with crabs in their tilting boxes were repeated at temperatures of 10, 15, 20 and 25°C their patterns of rhythmicity remained broadly similar (Naylor, 1963). Certainly their rhythms bore no relationship to environmental rhythms of local or residual geophysical origin that might not have been controlled in the laboratory, but it is not necessary to rely on indirect evidence such as this to support the notion of internal clock control. By rearing the larvae of *Carcinus maenas* from the egg to the juvenile crab stage, entirely in a day/night cycle, it was shown that young crabs with no experience of tides could be induced spontaneously to exhibit circatidal rhythmicity (Williams and Naylor, 1967). Initially after metamorphosis from the megalopa to the first crab stage artificially reared crabs showed only circadian rhythmicity, reflecting the day/night cycle in which they were reared. However, by exposing such juvenile crabs to a temperature of 4°C for a few hours and conveying no information concerning the 12.4-h periodicity of tides, it was possible to induce spontaneously into their behaviour pattern a walking activity rhythm of approximately tidal periodicity (Figure 2.5). This finding is best explained if the ability to express circatidal rhythmicity is an inherited characteristic of the crab's physiology, requiring only information about phase to set the tidal clockwork in motion (Williams and Naylor, 1967).

Further evidence of the fundamental nature of biorhythms in coastal, and more widely occurring marine organisms, is apparent in the rhythmic nature of several aspects of behaviour and physiology within any one organism. For example, four kinds of circadian rhythms are known to operate concurrently in the dinoflagellate *Lingulodinium* (*Gonyaulax*), the unicellular planktonic alga referred to above, which blooms periodically along the western coasts of northern USA. Rhythmicity in this alga is measured in laboratory cultures, which raises the question as to whether it is a phenomenon of populations or of

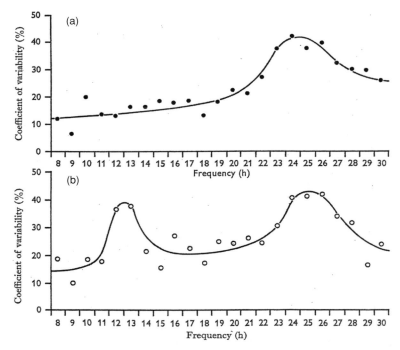

Figure 2.5 Periodogram analyses (a) of spontaneous locomotor activity pattern during four days in constant conditions of crabs reared in a 24 h light/dark regime and (b) of the same crabs chilled to 4°C for 15 h and maintained in constant conditions for 3 days thereafter (after Williams and Naylor, 1967).

individuals. There is evidence of weak coupling of circadian rhythms among cells in a laboratory culture, but most evidence points towards multiple circadian rhythms within single cells (Broda *et al.*, 1985). The four rhythms, all of different phasing, are of background luminescence (glow), which peaks at dawn, of bright flashing responses to mechanical stimulation, peaking at night, of photosynthesis, peaking around midday, and a rhythm of cell division which peaks sharply at dawn (Figure 2.6) (Hastings *et al.*, 1961; Kluge, 1982). Multiple rhythms are also seen in more complex organisms such as isopod crustaceans (Hastings, 1981b) and shore crabs, which exhibit circatidal rhythms of respiration and oxygen consumption, as well as locomotion, when kept in constant conditions in the laboratory. Crabs respire by taking water into their gill chambers situated on each side of the body. Water, carrying oxygen, enters the gill chambers through openings around the bases of the legs

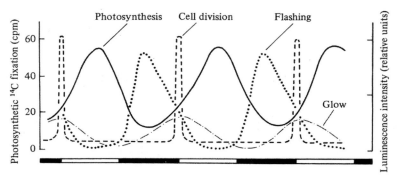

Figure 2.6 Diagrammatic representation of four concurrent circadian rhythms in the unicellular alga *Lingulodinium* (*Gonyaulax*). Bioluminescence occurs as background glow and as bright flashing in response to mechanical disturbance. (Abscissa shows 12 : 12 h pattern of light : dark experienced prior to transfer to constant continuous dim light.) (After Kluge, 1982; adapted from Hastings *et al.*, 1961.)

and is pumped forwards across the gills and mouthparts, to be expelled at the front of the body in exhalant swirls. The pattern of circulation is such that, with minimum disturbance to the crab it is easy to sample exhaled water by placing a rubber membrane over the anterior part of the body and attaching the membrane around a collecting tube. The volume of water pumped by the crab and its oxygen concentration before and after it passes through the gill chamber can then be measured over unit time at any time of day. Recordings have shown that individual crabs consistently pumped through their gill chambers twice the volume of water at the time of high tide that they did at low tide, even when tethered in an aquarium in constant conditions in the laboratory (Figure 2.7) and continued with that pattern for a few days (Arudpragasam and Naylor, 1964). Similarly, the spontaneous rate of oxygen consumption of the crabs increased by a factor of four or five at the expected times of high tide compared with low tide (Figure 2.7). Since the crabs were tethered the variations in respiration were not caused by increased locomotion at the time of expected high tide and therefore appeared to be under biological clock control. Other early examples of physiological phenomena of daily and tidal periodicity that were demonstrated to vary spontaneously include rhythms of colour change in a number of crab species. Among these, the pigment in the melanophores of the blue crab *Callinectes sapidus* and the fiddler crab *Uca pugnax* were shown to display a diurnal rhythm, dispersing during the day and concentrating at night, with a tidal rhythm of expansion and contraction superimposed upon it. Both rhythms in

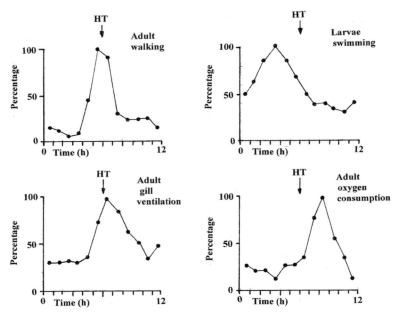

Figure 2.7 Clock-face readouts of the circatidal timing system of the shore crab *Carcinus maenas* kept in constant conditions in the laboratory. Values are expressed as percentages of the maximum value recorded during each expected tidal cycle for adult walking, gill ventilation volume and oxygen consumption, and larval swimming near sea bed. HT, time of expected high tide (redrawn from Naylor, 1985; Arudpragasam and Naylor, 1964; Zeng and Naylor, 1996b).

each of these North American crab species were shown to be endogenous, persisting under constant laboratory conditions (Brown *et al.*, 1953; Fingerman, 1955). A similar endogenous rhythm was demonstrated in the hatching process of larvae of the lobster *Homarus gammarus* by Ennis (1973). In normal light/dark cycles a female lobster releases larvae from the egg mass carried on its abdominal appendages for about one minute at nightfall and continues to do so on successive nights for two to six weeks afterwards. Evidence of endogeneity is provided since the nightly hatching rhythm continues for several nights in lobsters kept in continuous darkness. More recently, too, it has been shown that *Carcinus maenas* on the Atlantic coasts of Europe most frequently moults at the times of high tide, even in the laboratory (Abello *et al.*, 1997) and that newly caught females in laboratory aquaria release their newly hatched larvae at the times of nocturnal high tides (Zeng and Naylor, 1997). Moreover it has been reported that the clock-controlled hatching behaviour of the female crabs also sets the timing of circatidal swimming rhythms of their newly hatched larvae when they are released into the sea (Figure 2.7) (Zeng and Naylor, 1996a). Thus, not only adults, but larvae, too, behave as living clocks, adding to the medley of biological rhythms exhibited by coastal organisms and providing further evidence that tide-keeping ability, in addition to daily time-keeping capability, is an inherited feature of their physiology.

*   *   *   *

So far, most studies of 'tidal memories' in coastal organisms have been concerned with mobile species of the intertidal zone. Among these, even slow-moving limpets, such as *Patella vulgata*, which vacates its home scar to graze on films of encrusting algae on adjacent rock surfaces during low tides, has inbuilt circatidal clockwork that signals the time to return to the home location before the next high tide (Della Santina and Naylor, 1993). But, it seems, such body movements as are shown by otherwise immobile species that remain permanently attached to rocky substrates, are less likely to be under the control of internal timing systems. Such species include shore-living barnacles and mussels which open their shell valves and draw in water for feeding and respiration during high tide, but which close the valves again during low tide, reducing the risks of desiccation and predation by shore-feeding birds. Unlike mobile species, however, barnacles (Southward and Crisp, 1965; Sommer, 1972) and mussels (Ameyaw-Akumfi and Naylor, 1987a) appear not to persist with their tidal patterns of shell opening and closure when brought into the laboratory. For mobile species that forage underwater at high tide there would be a selective advantage for them to burrow or seek shelter and hide in crevices before they were exposed by the ebbing tide, particularly if they were vulnerable to desiccation and to predation by predators such as mammals and birds which feed on the shore at low tide (see Plate 5). On the other hand, for sessile species such as barnacles and mussels which are structurally well adapted to withstand desiccation and predation at low tide, there would be little adaptive value in the acquisition of anticipatory shell-closing behaviour as the tide ebbs. A purely exogenous closing response on exposure to air would suffice.

Similarly, for attached macroalgae in coastal localities there is little evidence of circatidal rhythmicity (Lüning, 2005). A great many of the rhythmic adaptive responses of plants, in general, have evolved in reaction to the day/night cycle, in view of their requirement for light during photosynthesis (Vince-Prue, 1982), and this also appears to be the case for marine intertidal algae. Among such algae endogenous rhythms, all of circadian periodicity, have been reported in a number of processes, including spore release (Andersson *et al.*, 1994), mitosis and growth (Makarov *et al.*, 1995; Titlyanov *et al.*, 1996) and chromatophore movements (Nultsch *et al.*, 1984). Also, some unicellular intertidal algae that undertake short vertical migrations from within the sediment on to the surface, do so mainly with circadian, rarely circatidal periodicity (see below), presumed to be phased by the day/night cycle (Palmer, 1976).

The apparent lack of tidal anticipatory behaviour in sessile animals such as barnacles and mussels is not, however, necessarily to say that they do not exhibit circatidal rhythmicity. The common mussel *Mytilus edulis*, for example, does so, as is apparent in its shell growth increments which, in intertidal mussels, are laid down like tree rings, but in tidal patterns which can be read quite clearly to determine the tidal history of the animals concerned. In order to do this, shells are first sectioned, polished and etched, after which transparent acetate peel replicas of the sections are made. Imprints of the growth rings on the transparent replicas can then be read off microscopically (Richardson, 1987, 1989). During spring tides, when low shore mussels are exposed to the air twice each day, a pattern of strongly defined tidal growth bands is apparent in the shells. Similarly patterned, but weaker, tidal growth bands also occur in the shells of low-shore mussels during neap tides when they are continuously submerged. Indeed, weak patterns of shell growth resulting from an endogenous rhythm of shell deposition are also apparent in permanently subtidal mussels, though with no apparent environmentally related periodicity. On the other hand, there are some seemingly sedentary animals between tidemarks that show not only tidal patterns of growth, but also endogenous circatidal rhythms of movement. These include many of the bivalve molluscs such as clams, cockles and their relatives that live just below the surface on sandy or muddy beaches. Many such species open their shells rhythmically in phase with high tides and extend into the water above the sand a pair of siphons through which they inhale and exhale seawater, extracting oxygen and filtering off suspended food particles as they do so, and they burrow deeper into the sediment at low tide. Some have been shown to continue with their tidally phased movements in constant conditions in the laboratory suggesting that they too possess endogenous tidal clocks (see Beentjes and Williams, 1986).

A related question to that posed with regard to sessile species concerns animals that live just below or just above tidemarks but which may at times be found intertidally, either at low tide or high tide. For example, on European beaches disturbance of the sand and stranded material just above the level of the highest tides usually reveals masses of sand hoppers, notably *Talitrus saltator* (Plate 6) and its relatives, amphipod crustaceans that, by day, normally lie hidden beneath tidally stranded debris or buried a few centimetres in the sand below. At night, large numbers of the amphipods can be seen by torchlight foraging for food on the sand surface, moving downshore if the tide is low, yet by dawn all are usually buried at the strand line again, unless disturbed by shore-feeding birds. This pattern of nocturnal foraging and daytime burrowing is repeated in constant conditions in the laboratory (Plate 7), indicative of the fact that the amphipods possess inbuilt circadian biological clocks (Figures 2.3, 3.4) (Bregazzi and Naylor, 1972). On the Atlantic coasts of Europe *Talitrus saltator* does not exhibit biorhythms of circatidal periodicity, suggesting that, even though they migrate down the shore when the tide is out, there is no biological requirement for them to do so. If it is dark, they evidently forage opportunistically wherever they are able to do so, even if the tide is in. Indeed, on Mediterranean beaches where tides are negligible, the same species of amphipod disperses away from the water's edge at night and forages inland (Mezzetti *et al.*, 1994). There is little adaptive requirement for *Talitrus*, living just above high tide mark, whether on tidal or non-tidal beaches, to have evolved tidal time-keeping ability.

Nor, it would seem, has there been intense selection pressure for some animals that live just below tidemarks to have evolved tidal timekeeping ability, even though they may experience tidal time-cues related to the rise and fall of tides above them or to lateral tidal flows. As in vertically migrating open ocean plankton species, discussed in Chapter 7, the dominant environmental signals against which a time sense has evolved in subtidal species in the coastal zone again seem to be those associated with the day/night cycle of light and dark. A good example of this is seen in two closely related species of swimming crabs commonly found around British coasts. One species, *Liocarcinus holsatus*, captured in the intertidal zone of a sandy beach showed a strong circatidal rhythm of locomotory activity in the laboratory, with maximum activity at the times of expected high tide. In contrast, the closely related *Liocarcinus depurator* captured from its normal habitat below tidemarks showed only circadian rhythmicity, with maximum activity during hours of expected daylight (Abello *et al.*, 1991). Another example, in this case of relevance to economically important fisheries, concerns *Nephrops norvegicus*, the Dublin Bay Prawn or Norway Lobster (Plates 8 and 9). Fishermen have long known that trawling for *Nephrops* in European waters is more successful at certain times of day than at others, the precise times of catch varying according to locality and depth of fishing. But it was not known until the fishery was investigated scientifically that variations in catches were in some way related to the intermittent burrowing habit of the prawns (Chapman and Rice, 1971). In constant conditions in the laboratory, the prawns exhibit a circadian rhythm with peak locomotor activity occurring during times of expected darkness (Figure 2.3) (Atkinson and Naylor, 1976; Naylor and Atkinson, 1976), paradoxically coincident with times of lowest catches in trawls in the Irish Sea (Figure 2.8). Evidently peak locomotor activity occurs at times when *Nephrops* actively maintains its burrow by excavating sediment and expelling it away from the burrow entrance, at which times they are particularly alert and able to take avoiding action in the event of an oncoming trawl. Here is an example of a case where knowledge of the circadian (and/or circatidal) biorhythms of a commercially important species is of relevance in developing management recommendations for fisheries of that species, as for example concerning the *Nephrops* fisheries at various depths in the Mediterranean Sea studied by Aguzzi *et al.* (2003).

There are clearly many examples of animals living just above tidemarks and others that live below the intertidal zone which exhibit biological rhythms of only circadian periodicity and show no behaviour that correlates with tides. Therefore, if intertidal animals in laboratory conditions are responsive to subtle environmental variables as drivers of their tidally correlated locomotor activity patterns, as has sometimes been suggested, then it is necessary to explain the absence of such responses in close relatives of shore species that live just above or below the range of tides. It seems that the most economical explanation is that tidal memories are evident only in organisms that are influenced by tides and for which there is some selective advantage to have evolved circatidal clockwork. All the evidence, so far, therefore suggests that the original hypothesis proposed in the 1950s for *Carcinus* is supported and that, pending further questions and experiments, one can proceed on the likely assumption that crabs, and other shore-living species, do have some form of tidal memory, the basis of which is an internal, inherited circatidal clock mechanism. It should be noted, however, that some organisms that appear to possess circatidal clocks, in fact, do not, as is the case with many unicellular algae that have been studied (Palmer, 1976). Intertidal sediments are often the habitat of large populations of such algae which, during daytime low tides, emerge on to the sediment surface in concentrations such as to change the colour of the sand or mud surface to that of their photosynthetic pigments. Between-times the algae migrate downwards into the sediment where they remain during high tide and at night. A series of kite diagrams of the vertical distribution of one such unicellular alga, *Euglena obtusata*, throughout a tidal cycle

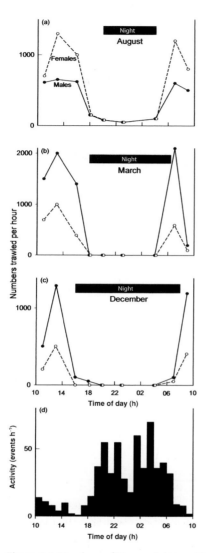

Figure 2.8 Numbers of Norway lobsters (*Nephrops norvegicus*) (closed circles, males; open circles, females) caught in standard trawl hauls at various times of day at 75 m depth in the north west Irish Sea in (a) August, (b) March and (c) December, plotted (d) against the endogenous circadian pattern of locomotor activity of freshly collected *Nephrops* recorded in constant conditions in the laboratory (after Naylor and Atkinson, 1976; with some data from Farmer (1974); by permission of Elsevier).

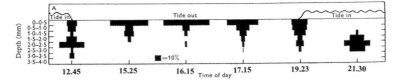

Figure 2.9 Kite diagrams of vertical distribution of *Euglena obtusa* in the surface layers of intertidal sediment during one natural tidal cycle. Solid blocks, per cent cells at indicated depths; wavy lines, high tide (after Palmer, 1976; modified from Palmer and Round, 1965; by permission of University of South Carolina Press).

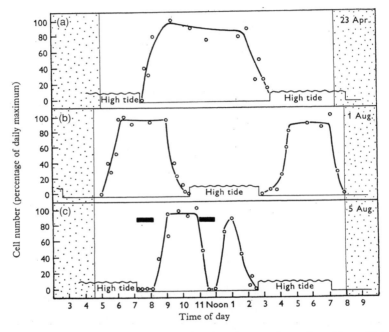

Figure 2.10 Vertical migration of *Euglena obtusa* to the surface of intertidal sediment in its natural habitat: (a) a summer day with a single daytime low tide; (b) a summer day with two daytime low tides; (c) experimental day with darkening of the substrate. Stippling, night-time; wavy line, high tide; black bar, artificial shading of exposed mudflat by canisters (after Palmer, 1976; modified from Palmer and Round, 1965; by permission of University of South Carolina Press).

is illustrated in Figure 2.9. When low tides occurred during the morning and evening of a long summer day, with high tide around noon, two peaks of emergence onto the sediment surface were apparent (Figure 2.10), suggesting that the pattern of emergence was tidal, driven by circatidal clockwork, as previously seen in some multicellular organisms such as crabs. However, the algae remained buried during night-time low tides and it was possible to suppress emergence during daytime low tides by artificially shading the sediment surface (Figure 2.10). In one species, *Hantzschia amphioxys*, it was demonstrated that the low tide pattern of emergence persisted in constant conditions in the laboratory, suggesting that it was under circatidal clock control but, in other species, upwards migration occurred only during the day. In a study of one chrysomonad species, 25 diatoms, three dinoflagellates and six euglenoids, all showed daytime vertical migration rhythms on the shore, but only ten of the species showed persistent vertical migrations in constant conditions in the laboratory, all of which were circadian in their rhythmicity. On the shore, the tides appear to shape the circadian rhythm of these photosynthetic unicellular algae into an apparently tidal form (Palmer, 1976). Even so, weak circatidal patterns of small-scale vertical migrations on the surface of intertidal sediments have been recorded in the diatoms *Hantzschia virgata* and *Pleurosigma angulatum* (Round and Palmer, 1966) and in the dinoflagellate *Amphidinium herdmaniae* by Eaton and Simpson (1979).

Quite early in the investigation of endogenous rhythmicity in coastal organisms it therefore became apparent that circadian and circatidal clockwork was involved and that preliminary hypotheses were required as to how such clockwork might function. A particular observation that it was necessary to accommodate in those hypotheses concerned the gradual loss of both circadian and circatidal rhythmicity expressed by organisms maintained in constant conditions in the laboratory. For example, if the crab *Carcinus* possessed one clock or separate individual biological clocks controlling circadian and circatidal rhythms one can visualize that these might run faster, slower or stop after a few days in the laboratory, but rhythmic walking activity expressed by *Carcinus* in actographs cannot be explained in this way. Successive peaks in a crab rhythm, prompted by the tidal memory, gradually subside and become less distinct, until eventually the pattern of behaviour is random. This gradual loss of behavioural rhythmicity is difficult to explain if each crab had only a single biological clock. However, if the internal clock system was composed of several clocks the observed behaviour of crabs in the laboratory could be more readily explained. The multiple clock model illustrated in Figure 2.11 postulates that the crab clock system is composed of a number of, say, cellular clocks, each of approximately 12.4-h (tidal) and approximately 24-h (daily) periodicity (Naylor, 1976). Some of the clocks might run at a slightly longer periodicity than tidal or daily and others slightly shorter. In nature, the clocks would be reset

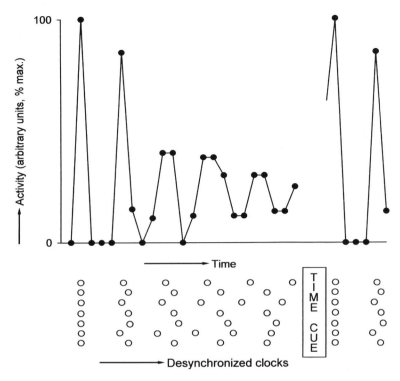

Figure 2.11 Hypothetical multiple clock model: clocks of approximately the same periodicity gradually drift out of phase as the expressed rhythm fades in constant conditions, to be reinstated when a time cue resynchronizes the clocks (redrawn from Naylor, 1976).

repeatedly by cyclical events in the environment but, under constant conditions in the laboratory, individual clocks or oscillators would drift out of phase from each other. Therefore the rhythm of behaviour expressed by a crab, at what might be called the clock-face, would be lost in the absence of external time signals in the form of appropriate environmental synchronizers. As the model implies, it should be possible to reinstate the rhythm of crabs that have become arrhythmic in the laboratory by exposing them to time signals to synchronize their multiple clocks and it will be seen in the next chapter that this can be done.

In trying to visualize the nature of biological clocks at this early stage, further observations that must be taken into account concern differences apparent in patterns of expression of clocks in animals from tidal localities when compared with those from terrestrial or semi-terrestrial habitats. Intertidal animals express behavioural rhythms that are 'noisy' and they often persist for only a few days in constant conditions, whereas the circadian rhythms of terrestrial animals more typically are expressed with greater precision and persistence. This may arise because, in contrast, despite the fundamental 12.4-h periodicity of tides, there are considerable regional differences in the pattern, timing and amplitude of tides. As discussed in Chapter 1, different localities around the British Isles, well within the geographical ranges of individual species of coastal animals, exhibit timings of high tide which vary by several hours. Within the gene pool of one species, individuals in one locality may experience tides exactly in antiphase to other individuals of the same species living elsewhere. Also, apart from regional differences there are often local differences in timing related to variations in atmospheric barometric pressure and associated weather conditions. Thus the 'noise' in the expression of tidal memories of intertidal animals might reasonably be considered as a corollary to the background of noisy rhythms in the intertidal environment. The process of adaptation seems to have compromised between persistent clock function on the one hand and, on the other, flexibility of responsiveness to local variability in the nature of tidal conditions. In fact, it has been hypothesized for some time that organisms can be considered to comprise suites of endogenous oscillators and that optimum performance of an organism's physiology depended upon the correct environmental synchronization of its various oscillatory processes (Pittendrigh, 1960, 1961). The hypothesis was confirmed in insects cultured in non-24-h environmental cycles which showed detrimental effects on growth and development, assumed to derive from internal desynchronization of their biological clocks (Aschoff *et al.*, 1971; Pittendrigh and Minnis, 1972; Saunders, 1972). In the cockroach *Leucophaea maderae*, non-circadian light/dark cycles also induce changes in the free-running period of behavioural rhythms and their responsiveness to light (Page and Barrett, 1987). Tests of the hypothesis concerning the possibility of perturbed growth in marine animals were carried out by Dalley (1979, 1980a), who compared the development and survival of the prawn *Palaemon elegans* in 24-h and non-24-h cycles of light and dark. In those experiments, prawns were reared from the egg, through zoea and post-larvae, to the juvenile stage in three types of light/dark cycle: 12 : 12 h L : D, 8 : 8 h L : D and random L : D. In all cases the prawns received the same amount of light in a 48-h period, were maintained at constant temperature and fed at irregular intervals throughout. Results indicated that growth was retarded and survival to the juvenile stage was significantly reduced in the 8 : 8 h L : D and random L : D conditions compared with the 12 : 12 h L : D controls

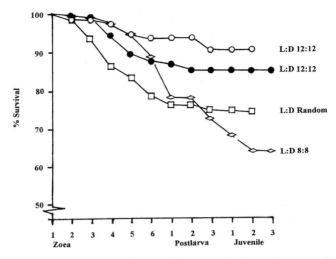

Figure 2.12 Percentage of larvae of the prawn *Palaemon elegans* surviving through six zoea stages and three post-larval stages to early juveniles during exposure to circadian (12 : 12 h L : D) and non-circadian (8 : 8 h L : D; random L : D) light regimes. A total of 40 to 50 larvae were used at the start of each experiment (redrawn from Dalley, 1979).

(Figure 2.12). In a similar study, using the brown shrimp *Crangon crangon*, survival, but not growth and morphological development, was adversely affected in non-circadian cycles (Dalley, 1980b). In *Palaemon*, the sex ratio was also shown to vary according to light/dark regimes, prawns cultured in 8 : 8 h L : D and random light/dark cycles being predominantly male (Dalley, 1979). Given these adverse effects on growth and survival in *Palaemon* and *Crangon*, assumed to be related to the mismatch between imposed environmental cycles and endogenous rhythmicity, the implications are clear for optimization of artificial culture procedures for marine crustaceans, and presumably for molluscs and fish, too (Naylor, 2005) (see also Chapter 6).

Finally, having acquired tidal time-keeping ability, are there other biological advantages that the innovation might confer? Certainly the possession of a time sense could help to explain how crabs and lobsters, which release their larvae intermittently, are able to do so at particularly advantageous times of day and state of tide (Ennis, 1973; Saigusa, 1982, 1986). It might also help explain the twice monthly appearance of swarms of midges at low spring tides at certain times of year, from larvae that develop from eggs laid between tidemarks by opportunist insects on some European shores (Neumann, 1987). Even more spectacularly, a time sense may also be involved in the annual spawning runs of the Californian grunion fish *Leuresthes tenuis* that rides the waves to spawn near the tops of beaches at the times of high spring tides (Clark, 1925) and the spawning of Palolo worms *Eunice viridis* on Samoan coral reefs at the third quarter of the moon in October or November (Caspers, 1984). Such events will be discussed in later chapters, seeking to establish whether they occur simply in response to environmental changes or to what extent they too might be driven by an internal time sense, in those cases of greater than circatidal or circadian periodicities.

But there is another type of behaviour for which a time sense is required and that is navigation in relation to the perceived position of astronomical features such as the sun and moon. The extent to which marine and coastal animals have evolved such sophisticated forms of behaviour, permitting them to be in the right place at the right time, will also be discussed in a later chapter. But first it is necessary to consider the mechanisms by which the tidal memories of coastal animals are imprinted and synchronized.

3

Tidal and daily time-cues

A clock is not much good if you can't pull it out of its stem and set it.
Arthur T. Winfree, 1987

On a hill near Liverpool, England, stands the Bidston Observatory. Built in 1867, it initially provided, by astronomical observation, the correct time for setting the chronometers of the ships that used the harbour and docks at the mouth of the River Mersey. With the advent of radio, the time-keeping function of the Observatory became redundant but, in keeping with tradition, it maintained responsibility for the timing of the 'One O'Clock Gun' that was a feature of Merseyside life until 1969. Earlier, however, in 1929, the building also took on an additional function when the University of Liverpool transferred into it ingenious facilities for the prediction of tides. Then renamed the Bidston Tidal Institute and Observatory, the facility became largely responsible for tidal predictions throughout the world in regions of British influence, helped on a smaller scale by a similar facility established under British jurisdiction in India. Even until late in the twentieth century there stood in the entrance hall of the Bidston building a mechanical tide predictor. It was an impressive museum piece after it ceased to be used in the 1950s, and was in the form of a large machine comprised of fixed and moving pulleys, gears and revolving drums. The pulleys were designed to mimic the changing interrelationships of the earth, moon and sun, in terms of their relative movements and distances from each other. When the machine was set up appropriately it plotted on a revolving drum the complete tide-curve for a given locality for any length of time ahead. It was the Doodson Tide Predictor, the last mechanical gadget to perform this function after Lord Kelvin first built such a machine in 1872. A personal computer will now do the same job in a much shorter time, but instruments built from pulleys, gears and drums for many years provided the data for tide-prediction tables throughout the world.

Foundations of the modern theory of tides were laid by Isaac Newton (1642–1727) when he applied his generalization of universal gravitation to the subject. Kepler (1571–1630) was among the first to formulate the idea that the moon exerted a gravitational pull on the water of the ocean, attracting it towards the location where the moon was overhead. He also explained that the water did not flow into the body of the moon because of the countervailing effect of the earth's gravitational pull. Newton's contribution was to show why there are two tides for each lunar transit, why the semilunar cycle of neap and spring tides occurs, and why equinoctial tides are usually larger than those at the solstices. Remarkably he also showed for localities around the world which experience only one tide during each lunar transit that the highest tides occur when the moon is farthest from the plane of the earth's equator (Pugh, 1987). The predictive power of Newton's gravitational theory as the basis for tidal science is evidenced by the availability of modern annual tide prediction tables that are essential reading for mariners worldwide and published, for example, by the United States Naval Oceanographic Office, the United Kingdom Hydrographic Office, and many other countries worldwide.

Tide tables cannot predict irregular changes in sea level that are attributable to meteorological events, but these are generally trivial and can be considered as 'noise' against the background of environmental tidal rhythmicity. So, with the repeatability of ocean tides, it is not surprising that some marine organisms have adapted to tidal rise and fall, given the strong signal/noise ratio of tidal oscillations in many localities throughout the world, as they have also adapted to the repeatability of the day/night cycle. Adaptation by natural selection is often difficult to demonstrate directly, since it is necessary to show that survival of the species is enhanced by the process involved. To prove that genetic adaptation to tides and day/night cycles has taken place during evolution it is necessary to demonstrate unequivocally that survivorship is higher in organisms that have body clocks than in related individuals or species that do not have such anticipatory skills. The scope and timescale of such experiments are serious limitations when working with large multicellular organisms and one awaits for comparison the discovery of naturally occurring individuals that by mutation have a time sense that differs from circadian or circatidal, or possess no time sense at all. Indeed, perhaps the first direct evidence that endogenous rhythmicity enhances fitness for survival derives from experiments with the prokaryotic cyanobacterium *Synechococcus*. Mutant strains of this unicellular organism have been discovered that expressed 'circadian' growth rhythms of 22, 25 and 30 h. In mixed cultures of the three strains, each grew best in an imposed light/dark cycle which approximated to the periodicity of its endogenous rhythm (Kando *et al.*, 1994; Ouyang *et al.*, 1998; Suzuki and Johnson, 2001). These results certainly suggest for prokaryotes that genetic adaptation to an external periodicity is of survival value.

The advantages of adaptation to environmental cycles which result in the acquisition of endogenous rhythmicity seemingly lie in an organism's endowed capability of anticipation of tidal or daily events. A circadian alarm system in a photosynthetic unicell would allow it to prepare its photosynthetic apparatus so as to capture the first photons at dawn (Suzuki and Johnson, 2001). It would also allow the phasing of incompatible processes to occur at different times, as in the case of photosynthesis and nitrogen fixation in cyanobacteria, since the nitrogen-fixing enzyme nitrogenase is inactivated in the presence of oxygen, which is released during photosynthesis (Mitsui *et al.*, 1986). Along similar lines, the 'escape from light' hypothesis proposes that the early evolution of circadian clocks could have been advantageous in phasing at night those cellular events that are inhibited by sunlight (Pittendrigh, 1993). Intuitively, too, a mobile animal's capability to anticipate tidal rise and fall seems likely to confer adaptive advantages concerning position maintenance between tidemarks and avoidance of predation and desiccation.

However, experimental evidence for the adaptive advantage of circadian and circatidal rhythmicity in large plants and animals in the coastal zone has yet to be established, bearing in mind the need for the discovery of clock mutants and the experimental difficulties concerning organisms with slow rates of reproduction. So far as coastal animals are concerned, the best that can usually be achieved is to elucidate as fully as possible the properties of an animal's biological clockwork and to determine how these relate to its lifestyle. By building up circumstantial evidence in this way one can suggest how responsiveness to tidal and daily cues takes place, and at least formulate the hypothesis that the possession of a biological clock is of adaptive value to coastal animals. Adaptation to the day/night cycle is easy to understand in that most animals possess eyes or more primitive light sensing organs that allow them to perceive and therefore respond to changes of light at dusk and dawn. Some terrestrial animals may even sense and respond to the changes in air temperature and humidity that are associated with the day/night cycle (Brady, 1979, 1982). However, the rise and fall of tides may impose on coastal animals a cacophony of environmental information, additional to that related to day and night. Tidal currents sweep and swirl across the shore, commonly at speeds of several knots. Waves driven by ocean swells and onshore winds crash on exposed rocky and sandy beaches with each high tide. With the rise and fall of tides, intertidal residents are not only pounded by wave action, but are also alternately exposed to the air and immersed in seawater, with associated problems of wetting and drying. In addition, tidal rise and fall exposes intertidal organisms to changes in water pressure. A number of localities around the world experience maximal tidal heights during equinoctial spring tides of 10 m or more, a head of water that induces an increase in hydrostatic pressure of an additional atmosphere above the ambient atmospheric pressure experienced, say, by animals at the lowest levels of the shore when they are exposed to the air at low tide. But, whatever the tidal amplitude, coastal animals are clearly exposed to cyclical changes of external pressure to which they might in some way be expected to respond. Moreover, associated with tidal rise and fall, it may also be the case that animals between tidemarks are exposed to tidally related changes of temperature. Rarely are sea and air temperatures identical at any one time, animals of temperate beaches are usually cooled by tidal immersion in summer but often warmed when immersed by the flooding tide during cold winters. Also, many intertidal animals will repeatedly be subjected to physiological shocks associated with alternating immersion in freshened and salt water. Particularly susceptible to the osmotic hazards of such conditions are organisms that live on estuarine shores, in pools on open beaches where freshwater streams spill down the shore, and even on apparently fully marine shores when they are affected by heavy downpours of rain at low tide. In the previous chapter, it was concluded that many coastal animals have tidal memories, based upon their repeated experiences of tidal rise and fall. Now one can postulate a number of possible tidal time cues, zeitgebers or environmental entraining signals that can be investigated experimentally to determine how intertidal animals might synchronize their biological clocks. The precise timing of human daily affairs has long been well served by events such as the city of Liverpool's One O'Clock Gun and more generally by radio time signals, but the question remains as to which of the many tidal variables enable coastal animals to synchronize their biological clocks.

*   *   *   *

Simple experiments that can be performed to gain preliminary information about possible tidal zeitgebers require the use of suitably versatile animals. Some species of shore crabs are ideal for this purpose since they survive well in the laboratory for a few days until they have had no recent experience of tides and can be expected to have at least partially lost their tidal memories. These can then be placed back on the shore from which they were originally collected, simulating the pioneering experiments with *Convoluta roscoffensis* carried out by Bohn (1903). With hardy and extensively mobile crabs, however, it is possible to consider returning them to different locations on the shore that vary according to a range of physical variables that the crabs might encounter during their normal travels. Some might be placed at the water's edge in tethered cages designed to float up and down with the tide. Over each subsequent tidal cycle such animals would be exposed to wave action and tidal currents, to only minor changes in hydrostatic pressure and temperature, and would not be exposed to the air at low tide. Others might be placed in cages that are firmly attached to the substratum just below the level of low tide. These animals would be subjected to wave action and tidal currents, and to marked changes in hydrostatic pressure as the tide rose and fell, but they would experience only minor changes in water temperature and would not be exposed to the air at low tide. Yet others might be anchored in cages at mid-tidal level on the shore, where they would be exposed to the full range of tidal cues: namely, to cycles of immersion and emersion, air and water temperature differentials, hydrostatic water pressure changes, and tidal changes in wave action and water currents. Such experiments were carried out using the shore crab *Carcinus maenas* from a strongly tidal locality on the coast of South Wales (Williams and Naylor, 1969). Crabs which had been isolated naturally from tides in a non-tidal harbour, and which normally exhibit only a circadian pattern of behaviour, were placed in weighted cages fixed to the substratum below tidemarks and other crabs were placed in cages fixed to rocks at mid-tide level. The cages with their resident crabs remained in place throughout several tidal cycles, after which the crabs were recovered and tested in constant conditions in the laboratory to determine whether they had refreshed their tidal memories. After 11 days (Figure 3.1), and even after only two days (four tidal cycles) (Figure 3.2) in their cages on the shore, both sets of crabs exhibited increased walking activity in the laboratory at the expected times of high tide on the shore outside, with alternate peaks indicating retention of the circadian component of their behaviour. The results suggested that the crabs from a non-tidal harbour had not lost their circatidal clockwork, nor had they lost their ability to utilize environmental periodicities to synchronize that clockwork. Beyond that, however, it was also clear that the crabs that had been anchored at mid-tide level on the shore (Figure 3.1) showed greater precision and persistence of their tidal memories than others

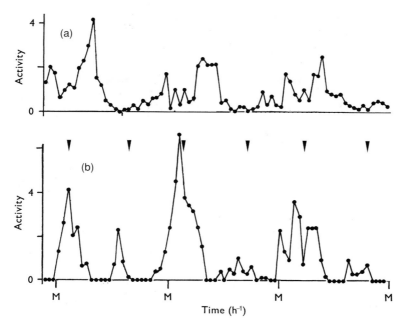

Figure 3.1 Mean hourly locomotor activity of crabs (*Carcinus*) from a non-tidal harbour recorded under constant conditions in the laboratory: (a) 12 freshly collected controls; (b) eight crabs after 11 days in a creel fixed on a tidal shore at mid-tide level. Activity expressed as mean actograph tilts/h; M, expected midnight; arrows, expected times of high tide (after Williams and Naylor, 1969).

kept in cages below low water mark (Figure 3.2). This suggested that shore crabs probably do not rely on one tidal timer but use a number of environmental cues to set their tidal clocks. Since it could be assumed that the crabs in cages were relatively sheltered from water movements, those below tidemarks were perhaps using changes in hydrostatic pressure to synchronize their clocks. In contrast the improved tide-keeping of the crabs entrained between tidemarks indicated that they were evidently using not only hydrostatic pressure changes, but other variables too, associated with alternating periods of exposure to air and cycles of temperature change (Williams and Naylor, 1969).

On the basis of simple field experiments such as these, it was then necessary to test whether each possible tidal time cue could be shown to function individually in that capacity in the laboratory. Testing for the effects of temperature cycles proved to be the simplest experiments to carry out. Crabs were kept individually in rocker box actographs to

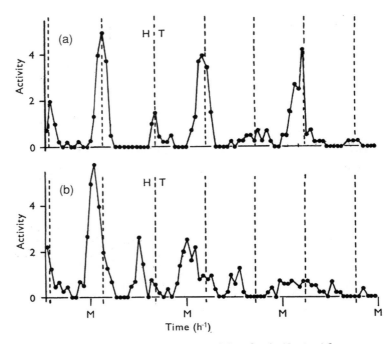

Figure 3.2 Mean hourly locomotor activity of crabs (*Carcinus*) from a
non-tidal harbour recorded under constant conditions in the laboratory:
(a) five crabs after 2 days in a creel fixed on a tidal shore at mid-tide level;
(b) four crabs after 2 days in a weighted creel placed below low water
mark. Activity expressed as mean actograph tilts/h; M, expected midnight;
vertical dotted lines indicate expected times of high tide (after Williams
and Naylor, 1969).

record their walking activity, and the actographs were transferred from
one constant temperature room to another at a different temperature
with the minimum of disturbance. In one such experiment crabs were
exposed to alternating intervals of 6.2 h at 24°C and 13°C for a total of
five days, equivalent to ten artificial tidal cycles of temperature alone.
Locomotor activity of the crabs was recorded continuously throughout
the treatment days, and continued afterwards when the crabs were kept
at a constant temperature of 13°C. During the treatment stage of the
experiments crabs did not move when the actographs were transferred
from one room to another, nor did they show increased locomotor activity during the expected times of high tide. They showed marked increases
in locomotor activity only when exposed to 13°C and remained quiescent
at the higher temperature (Figure 3.3). The lower temperature would be
that of the sea in June at the South Wales locality where the crabs were

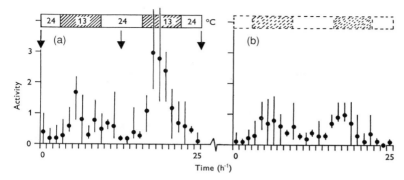

Figure 3.3 Hourly locomotor activity values of five crabs (*Carcinus*) freshly collected from a tidal shore and exposed in the laboratory to (a) an artificial tidal cycle of temperature change for 5 days, followed by (b) 3 days in constant conditions. Values are expressed as mean and range, summed over standard 25 h (approximately twice-tidal) intervals; arrows indicate times of expected high tide; shaded areas in post-treatment indicate times of expected low temperature (after Williams and Naylor, 1969).

collected, so their repeated bursts of walking activity at that temperature in the laboratory appeared to reflect crab movements on the shore at high tide. A temperature of 24°C would be an exceptionally high air temperature at the collecting site. However, air temperatures are consistently several degrees higher than sea temperatures from March to September where the crabs were collected, suggesting that exposure to high air temperatures might reinforce the suppression of locomotor activity of shore crabs when they normally secrete themselves beneath rocks at low tide (Williams and Naylor, 1969).

A follow-up experiment was clearly indicated, to expose crabs to a cycle of temperature change with a differential of 4°C that more realistically reflected the difference between mean sea and air temperatures in summer along the coasts of South Wales. In such an experiment, crab movements were certainly subdued during 6.2 h episodes at 17°C and stimulated during intervening 6.2 h sessions at 13°C. However, after five days (ten 'tidal' cycles) of such treatment, crabs kept subsequently at 13°C showed no evidence of persistent circatidal rhythmicity. This suggested that clock-setting with a 4°C differential alone requires longer than five days for success, or that additional time cues are necessary for quick resetting of the clocks. Indeed, successful resetting of crab clocks was achieved in five days in animals exposed to alternating sessions of 6.2 h at 13°C in seawater and 6.2 h at 17°C in air. Shore crabs therefore appear

to respond to combined environmental cycles of temperature and immer-
sion/emersion in the process of quick synchronization of their biological
clockwork (Williams and Naylor, 1969).

In view of the apparent added value of periodic tidal immersion
in enhancing temperature synchronization of crab clocks, it was subsequently possible to test for that variable alone as a tidal time cue for
shore crabs. Using a computer driven tide machine that filled and emptied a tank in which crabs were kept, movements of the crabs were monitored continuously (Reid et al., 1992). As is often the case in biological
experiments, there was some variability in the results but, in this case,
explanation of the variability turned out to be straightforward when the
lifestyles of crabs of various sizes were taken into consideration. Generally, the largest crabs seemed not to be responsive to immersion/emersion
cycles alone as time cues for their biological clocks. These are members
of the shore crab population that, by and large, migrate on and off the
shore with each tide. Some may be stranded in low shore rock pools at
low tide but they are relatively infrequently stranded in air on the ebbing
tide. In contrast, recently settled juvenile crabs generally spend all their
time fairly high in the intertidal zone and it is only such crabs, not large
adult males and females, which readily synchronize their tidal clocks to
artificial cycles of wetting and drying in the laboratory (Reid et al., 1992).

The question then arises as to how environmental cycles induce
biorhythms. Shore crabs exposed to several cycles of appropriate tidal
variables quite quickly acquire a persistent, that is circatidal, rhythm
that matches the imposed artificial tide, but the results could suggest that
the crabs have in some way remembered the period of the imposed tidal
cycle. Some evidence that they possess a true tidal clock was discussed
in Chapter 2, which showed that tidal rhythmicity can be induced in
previously arrhythmic crabs by a short exposure to low temperature
(see Figure 2.5). However, the question remained as to whether tidal
clock adjustment could be shown to take place in response to a short
tidal time cue in a normally tidally rhythmic animal. Certainly it is well
established that the phase of circadian rhythms can be readily shifted by
a short pulse of an environmental variable (Harker, 1964), as occurs with
the circadian rhythm of walking activity expressed by the sand hopper
Talitrus saltator (Bregazzi and Naylor, 1972) (Figure 3.4). Moreover it is also
well known that timing of a circadian rhythm can be delayed or advanced
depending upon the time of day when a normally rhythmic animal is
exposed to an environmental time cue. Such phase responsiveness of the
circadian locomotor rhythm of *Talitrus* when exposed to 2 h pulses of
white light of 400-lux intensity applied at various times of day and night

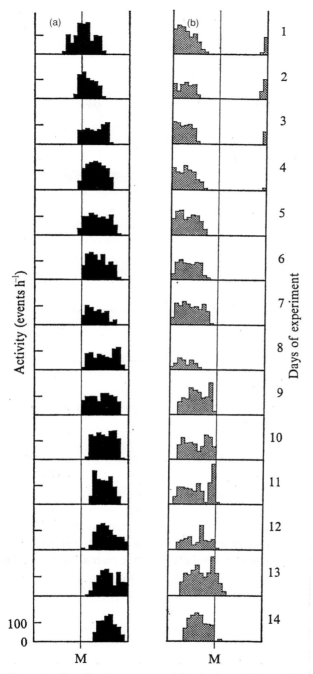

Figure 3.4 Circadian locomotor activity rhythms of the sand hopper
Talitrus saltator maintained in the laboratory for 14 days at constant
temperature and in continuous dim red illumination: (a) six freshly
collected animals; (b) six freshly collected animals chilled to 3°C for 12 h.
M, expected midnight (after Bregazzi and Naylor, 1972).

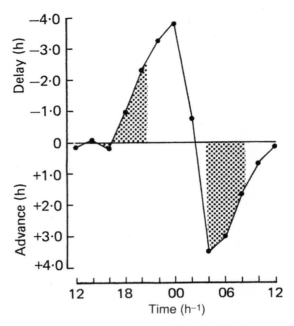

Figure 3.5 Phase response curve of the circadian locomotor activity rhythm of *Talitrus saltator*, indicating that 2 h pulses of light before expected midnight induce delays in subsequent peaks and 2 h pulses applied after expected midnight induce phase advances (after Williams, 1980).

is illustrated in Figure 3.5. Light pulses applied from dusk to midnight result in subsequent delays in the phase of the sand hopper rhythm, whilst pulses applied during the early morning result in phase advances of the normal nocturnal bursts of activity. The switching point from rhythm delays to rhythm advances occurred around the time of midnight (Williams, 1980).

To try to resolve the question of phase responsiveness for a tidal rhythm the clock responses of a New Zealand rocky shore crab *Hemigrapsus edwardsi* were studied in response to a single artificial tidal time cue applied at different times of the environmental tidal cycle (Naylor and Williams, 1984). At the Portobello Marine Laboratory of the University of Dunedin, it was first established that the crabs exhibited a circatidal rhythm of walking activity when kept in moist air at constant temperature in the laboratory, reflecting their behaviour on the shore. Then, using freshly collected crabs recorded in constant conditions in moist air in the laboratory, the crabs were disturbed only once when they were immersed for 3 h in cold sea water, simulating a short exposure to high tide, before returning them to the original moist air conditions. Different crabs were exposed to the three hour 'tide' at different phases of their naturally running tidal clocks, in order to assess how their tidal clocks responded, if at all, to the tidal time cue. In most cases the pattern of expression of the *Hemigrapsus* tidal rhythm was unaffected by exposure to the single 'tide' and the rhythms ran as normal, peak walking activity coinciding with the times of high tide on the shores outside the laboratory where the crabs were collected. However, it was found that the rhythms showed phase changes if the tide cue was initiated during a 4-h interval around the time when the crabs would have expected to experience high tide. If the crabs were immersed 2–3 h after they would normally have been covered by the rising tide, their subsequent bursts of walking activity were slightly delayed. In contrast, if the crabs were immersed for 3 h, starting about 4 h after they would normally have been covered by the rising tide, their response was quite different. It was as if the crabs were responding to the next high tide and the timing of their later bursts of circatidal walking activity was advanced accordingly. These results suggested that the crab clocks were susceptible to adjustment around the times when they expected high tide to occur, but not at other times during the tidal cycle. They were responsive to immersion by the artificial tide provided it occurred during a critical 4-h phase centred on the time when they expected to experience high tide (Figure 3.6). This phase responsiveness suggests that in real life on the beach the biological clocks of the crab *Hemigrapsus* are synchronized by the process of immersion during each flood tide.

Similar tidal patterns of phase responsiveness have been found in a number of intertidal animals that are known to possess circatidal clocks. These include the gastropod mollusc *Littorina* (Petpiroon and Morgan, 1983), an amphipod crustacean *Corophium* (Harris and Morgan, 1984a) and the fish *Lipophrys* (Northcott *et al.*, 1991), each of which responded to a single pulse of an environmental stimulus that the animals normally associate with exposure to high tide. In each case, the response was similar to that seen in the crab *Hemigrapsus,* with phase delays and phase advances of their circatidal rhythms occurring when the tidal pulses were applied just before and just after the times when the animals expected to experience high tide. Such patterns of phase responsiveness contrast sharply with those of terrestrial animals, including those of animals that live just above tidemarks such as the sand hopper *Talitrus saltator.* In those animals the timescale of phase responsiveness is 24 h, not the tidal time base of about 12 h. In terrestrial animals, inhabiting largely 24-h environments,

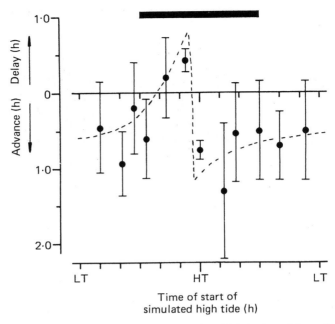

Figure 3.6 Phase response curve of the circatidal rhythm of the crab
Hemigrapsus edwardsi, indicating that 3-h pulses of simulated high tide
before expected high tide induce phase delays in subsequent peaks and 3 h
pulses of simulated high tide after expected high tide induce advances
(after Naylor and Williams, 1984).

particular scientific interest has focussed on the phase responsiveness of
circadian clocks to pulses of light, and on the entrainment of circadian
rhythms by cycles of light and dark. However, coastal animals are exposed
to a wider variety of cyclic environmental signals than those experienced
by terrestrial animals. The question remains, therefore, whether environmental cycles other than temperature change and tidal immersion are
involved in the entrainment of tidal rhythms in coastal zone species.

Among the other environmental variables experienced by intertidal animals, the clock-setting role of changes in hydrostatic pressure
induced by the rise and fall of tides was investigated by Jones and Naylor (1970) working with the sand beach isopod crustacean *Eurydice pulchra*
(Plate 10), whose sensitivity to changes in water pressure was well known.
Studies by Morgan (1965) and Knight-Jones and Morgan (1966) had previously shown that a number of crustaceans including *Eurydice*, when
kept in seawater in a pressure chamber, could be induced to swim by

increasing the hydrostatic pressure and to stop swimming and sink passively if the pressure was reduced. A group of *Eurydice* maintained in a similar pressure chamber was therefore subjected to an artificial tidal cycle of pressure change from ambient atmospheric pressure to an additional 0.5 atm, equivalent to a depth increase of 5 m of water, every 'high tide', for five days (Jones and Naylor, 1970). As expected, the isopods swam vigorously every time the pressure was raised, but the question was how the animals would behave when pressure cycling was discontinued. At constant atmospheric pressure after entrainment the isopods persisted with their rhythmic pattern of swimming, their most energetic bursts occurring at the times when they 'expected' hydrostatic pressure to increase. The outcome of the experiment suggested very strongly that the tidal clocks of *Eurydice* had been reset and that water pressure changes associated with tidal rise and fall serve as a natural zeitgeber for the swimming rhythms of these isopods in their normal beach environment.

Later more extensive experiments were carried out on the synchronization of tidal clockwork by hydrostatic pressure using the common shore crab *Carcinus maenas* (Naylor and Atkinson, 1972). Those experiments began with the construction of clear plastic pressure chambers that could be filled with seawater and within which the movements of crabs could be monitored electronically as they moved through a beam of infra-red light projected across each chamber (Plate 11). Initial experiments established that the crabs were responsive to hydrostatic pressure changes and were sensitive to pressure changes within the range of pressure variation that they would normally experience between low tide and high tide. Then, further experiments were undertaken to find out if the tidal clocks of crabs could be reset by exposing them to a range of artificial tidal cycles of hydrostatic pressure, simulating the magnitude and pattern of pressure changes that would be experienced in their normal habitat. Artificial tidal cycles varying from ambient atmospheric pressure to an additional 0.1, 0.2, 0.3 and 0.6 atm all generated locomotor activity at the raised pressure intervals of the four entrainment regimes. Moreover, at continuous atmospheric pressure after six days' treatment, all crabs continued to show increased walking activity at times when they would expect pressure to increase if the entrainment regime had continued (Figure 3.7). It could be concluded that hydrostatic pressure cycles generated by tides covering crabs by 1, 2, 3 or 6 m of water, well within the normal range of tides where the crabs were collected, are able to serve as efficient time cues to synchronize the biological clocks of shore crabs. Also, by carrying out similar experiments on crabs throughout the year, it was found that crabs were most responsive to the tidal

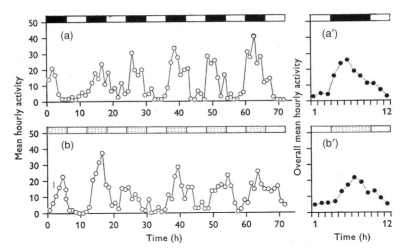

Figure 3.7 Mean hourly locomotor activity of six previously arrhythmic crabs *Carcinus maenas* recorded (a) during 72 h exposure to alternate approximately 6-h episodes of atmospheric pressure (ambient 1.0 atm) and an additional 0.6 atm, equivalent to a high tide of 6 m, followed immediately by (b) 72 h at constant atmospheric pressure. Activity is expressed as the number of times each hour that crabs interrupted infra-red light beams shone across the recording chambers. Black bars, episodes of raised pressure; stippled bars, expected episodes of raised pressure (after Naylor and Atkinson, 1972).

time cue of hydrostatic pressure during late summer, and less so at other times of year (Naylor and Atkinson, 1972). Late summer is the time of year in temperate latitudes when sea temperatures are at their maximum and air temperatures are beginning to fall. At this time of year, the differences between sea and air temperatures are at their smallest and therefore at their least effective as tidal time cues. In short, there is seasonality in the extent to which different environmental tide cues are used as synchronizers of crab biological clocks, hydrostatic pressure cycles becoming most significant when temperature cycles are least effective in that role.

Yet another possible tide cue for shore crab clocks, namely that of salinity, was investigated by monitoring crab rhythmic behaviour in a recording chamber which incorporated an electronically controlled pumping system that periodically flooded the crabs with water of different salt concentrations. In this way the crabs would, for predetermined intervals of time, find themselves immersed in seawater, in freshened seawater or in concentrated seawater to permit measurements to be made of their locomotor activity as they moved between infra-red emitting

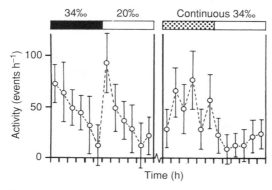

Figure 3.8 Hourly locomotor values of five crabs (*Carcinus*) freshly collected
from a tidal shore and exposed in the laboratory to an artificial tidal cycle
(6 : 6 h) of salinity change from full seawater (34‰) to dilute seawater
(20‰) for 36 h, followed by 36 h at constant salinity of full seawater.
Activity on the vertical axis is expressed as mean and standard deviations,
summed over standard tidal (approximately 12 h) intervals; stippled bar,
time of expected full seawater at constant salinity after entrainment (after
Taylor and Naylor, 1977).

diodes and receptors fitted on opposite sides of the transparent recording chambers (Taylor and Naylor, 1977). Initially crabs were exposed
to artificial tidal cycles of salinity change when for half of the cycle
(6.2 h) they experienced normal seawater and, for the other half, to seawater diluted with fresh water to two-thirds normal strength, simulating
the influx of freshwater to a rock pool at low tide. Walking activity of
a number of the crabs was recorded continuously during six artificial
tidal cycles of changing salinity, and then for a further 36 h, equivalent
to three tidal cycles, at constant normal salinity. In constant conditions
after treatment the crabs showed a clear and persistent circatidal rhythm
of walking activity, with bursts coinciding more or less with the times
when the crabs would have 'expected' normal seawater conditions if the
treatment regime had continued (Figure 3.8). Similar refreshing of the
crab's tidal memories occurred when the imposed artificial tidal cycles
of salinity varied from that of normal seawater to 150% seawater, concentrated by the addition of sea salt to mimic water evaporation from an
isolated rock pool at low tide. Such treatment again stimulated crabs into
walking activity when they experienced the episodes of normal seawater,
in a pattern that persisted as a circatidal rhythm afterwards, whether the
crabs were maintained subsequently in normal, dilute or concentrated
seawater at constant levels (Taylor and Naylor, 1977).

There was, however, a strikingly different outcome from the experiments involving entrainment of tidal rhythms by salinity changes when compared with those using entrainment cycles of temperature or pressure. During treatment with artificial tidal cycles of temperature and hydrostatic pressure, crab activity was stimulated only during the 'high tide' phase of the entraining cycle when they were exposed to low temperature and high pressure. In contrast, during exposure to artificial tidal cycles of salinity change, crabs responded by being active not only on exposure to normal seawater, that is the high tide condition, but also to the alternating increases or decreases in salinity, equivalent to low tide in the entrainment regime (Figure 3.8). However, though the rhythmic pattern of response to simulated high tide episodes of normal seawater persisted in constant conditions after entrainment, the responses to the low tide conditions during entrainment did not continue in constant conditions afterwards. Therefore, though the responses to the high tide conditions of normal seawater did serve as tidal time cues for the biological clocks of crabs, the responses to low tide conditions of dilute or concentrated seawater did not. It is presumably the repeatability and predictability of the sea's normal salinity at high tide as an effective tidal time cue that has been selected for during evolution. Other salinities would present themselves more sporadically to crabs that move around between tidemarks, dependent upon rainfall and other freshwater runoff, or evaporation from rock pools at low tide in summer. Such sporadic events would be less suitable time cues than consistently repeated immersion in seawater at high tide, so that over evolutionary time they have remained as simple exogenous responses to external variables without being coupled to the endogenous clockwork of the crabs. That is not to say, however, that the responses of crabs to unusual salinities are of no significance. The exogenous responses of green crabs to increased or decreased salinities can certainly be considered to enhance their chances of survival. At low tides crabs may find themselves in pools where salinity decreases by freshwater inflow, or increases by evaporation, and escape reactions stimulated by the changes in salt concentration would increase the chances of crabs moving to a more favourable salinity regime nearby. Such escape reactions stimulated by the abnormal salt concentrations of their surrounding water can be regarded as behavioural traits that enhance a crab's ability to react to unfavourable osmotic surroundings, over and above its well-established capability to regulate the concentration of its body fluids (Thomas et al., 1981).

Evidence so far, therefore, indicates that shore crabs are not only able to respond to simulated cycles of hydrostatic pressure, temperature

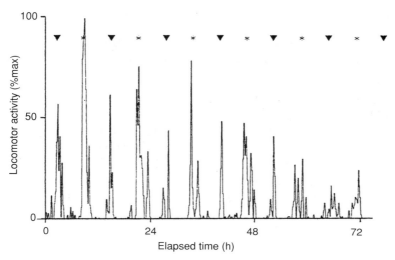

Figure 3.9 Entrainment of a circasemitidal (approximately 6 h) locomotor rhythm in *Carcinus maenas* after 4 days' exposure to artificial tidal cycles of hydrostatic pressure and salinity applied in antiphase. Arrows, expected time of maximum pressure; asterisks, expected times of maximum salinity (after Reid and Naylor, 1990).

and salinity, but are also able to adjust their clocks in the process of those responses. This raises a new question as to whether there is only one component of crab clockwork that responds to any or all of the tide cues to which they are exposed, or whether there are multiple clock-setting mechanisms. Using more sophisticated environmental simulators than used hitherto, that question was addressed by exposing crabs to artificial tidal cycles of two or three variables which could be imposed out of phase with each other. When cycles of hydrostatic pressure and salinity were imposed in antiphase, such that crabs experienced alternating 'high tide' conditions of high pressure and high salinity at approximately 6-h intervals, they showed increased locomotor activity during each 'high tide' and continued to express a free-running semitidal rhythm in constant conditions afterwards (Figure 3.9) (Reid and Naylor, 1990). Similarly, when exposed to artificial tidal cycles of the three variables of pressure, temperature and salinity applied together, but approximately 4 h out of phase with each other, crabs again recognized each 'high tide' cue. In those experiments many of the crabs tested showed bursts of locomotor activity at approximately 4-h intervals, and continued to do so in constant conditions afterwards (Warman and Naylor, 1995). So, shore crabs are able to adjust their clocks to permit them to anticipate two, or even three, high tide events that occur within the timescale of a normal tidal cycle, supporting the notion that each crab has multiple clock components that respond to tidal time cues, probably with different components cued by specific tidal variables.

Another tidal variable that has been shown to act as a tide cue that entrains circatidal rhythms is disturbance by wave action, which can be simulated in the laboratory by artificial tidal cycles of mechanical agitation. This mechanism of entrainment is particularly evident in sand beach-living species that spend part of their lives buried in intertidal sand when the beach is exposed to the air at low tide and another part swimming in the water covering the beach at high tide after their zone on the beach has been subjected to wave action during the rising tide. The first unequivocal evidence for mechanical disturbance as a tidal time cue was provided by Enright (1965b) who studied the sand-beach isopod crustacean *Excirolana chiltoni*. In nature, this swimming isopod burrows into sand at low tide, emerging to forage in the surf as the tide rises over its native beaches in California (Enright, 1965b, 1976a, b). When brought into constant conditions in the laboratory the animal persists with its tidally related rhythm of burrowing and swimming indicative of the fact that it possesses a circatidal clock. Then, pioneeringly, Enright demonstrated that the phase of the circatidal rhythm of swimming by *Excirolana* could be reset by exposing the isopods to artificial tidal cycles of stirring to simulate wave action at 'high tide' alternating with episodes of quiescence to simulate low tide. Use of tidal agitation as a time cue for the setting of tidal rhythms is also apparent in *Eurydice pulchra,* a close relative of *Excirolana* that lives on European sandy beaches (Jones and Naylor, 1970; Hastings, 1981a), though resetting may be diurnally modulated depending upon the feeding state of the isopods at the time of entrainment (Reid, 1988). In addition Neumann (1976a) reported entrainment of tidal rhythmicity by simulated tidal action in two species of East African fiddler crabs, *Uca urvillei* and *U. annulipes,* in a sand hopper *Talorchestia quoyana* and in the flatworm *Convoluta roscoffensis.*

With its extensive swimming behaviour at high tide and its occupation of a precise zone, burrowed in the beach at low tide, it seemed a reasonable assumption that *Eurydice pulchra* would, like its relative *Excirolana*, when brought into the laboratory, exhibit a spontaneous circatidal rhythm of swimming and quiescence. This certainly proved to be so when the swimming behaviour of freshly caught specimens was recorded in aquarium tanks provided the tanks did not contain sand into which the isopods could burrow (Figure 2.3) (Jones and Naylor, 1970). Surprisingly, however, when a layer of clean sand was introduced at the bottom of the aquaria, into which the isopods were able to burrow, in some, but not all experiments was it found that they emerged in large numbers to swim at the times of expected high tide. Paradoxically they appeared to have a perfectly good tidal memory but, in laboratory tanks, many sometimes appeared to ignore their internal tidal clock and did not emerge from sand when they might have expected to experience high tide. On the beach the isopods evidently rely heavily upon wave action during the rising tide to induce them to emerge from the sand and swim; only occasionally, it seemed, did they appear to rely on their tidal clockwork to time their emergence as the tide rises. Importantly, the onset of wave action appears also to serve as a time-cue that reinforces the phasing of the isopods' circatidal clockwork, the critical and consistent role of which is to time the end of the swimming session at high tide, thus ensuring that they re-burrow in sand between tidemarks and avoid being carried to unfavourable localities offshore (Jones and Naylor, 1970). Even so, it remained to explain the behaviour of *Eurydice pulchra* in the laboratory whereby in some experiments, but not in others, they emerged spontaneously from sand to swim at the times of expected high tide. These apparently contradictory findings were explained only when it was discovered that spontaneous emergence at high tide by *Eurydice pulchra* tended to occur around the times of spring tides and not at neaps (Alheit and Naylor, 1976), indicative of seemingly moon-related behaviour discussed in Chapter 6.

The role of light in controlling the rhythmic behaviour of coastal animals is, perhaps as expected, primarily related to the synchronization of circadian components of behaviour. Certainly, high shore animals like the strandline sand hopper *Talitrus saltator*, and subtidal forms such as the Norway lobster *Nephrops norvegicus*, show only circadian rhythms of movement (Figure 2.3) that are primarily entrained by day/night cycles of light and dark. When freshly captured *Talitrus* are brought into constant conditions in the laboratory they initially emerge spontaneously from sand and forage on the surface for a few hours around the time of 'expected' midnight. On the next night the pattern is repeated, though with a slightly later time of emergence, and the rhythm of emergence and foraging continues on subsequent nights at slightly longer than 24-h intervals, in typical circadian fashion (Figure 3.4) (Bregazzi and Naylor, 1972). The Norway lobster (*Nephrops norvegicus*) also exhibits fairly typical circadian rhythmicity when maintained in constant conditions in the laboratory (Atkinson and Naylor, 1976), a rhythm which is highly sensitive to the daily pattern and quality of dim light changes that occurs at depth in the sea, as was seen in Chapter 2. In the Mediterranean Sea there is evidence for *N. norvegicus* from depths down to 400 m that blue light of 480 nm wavelength is the putative zeitgeber for their emergence rhythms which are sometimes atypical, possibly cued by Internal Currents (Aguzzi and Sarda, 2008) and that rhythmicity of catches of the burrowing shrimp *Processa canaliculata* from the same localities is also phased to optimum light intensities (Aguzzi *et al.*, 2008).

* * * *

Further questions can now be raised concerning coastal animals that show both circatidal and circadian rhythms of behaviour when kept in controlled conditions in the laboratory. How would they behave if they lived permanently away from the influence of tides? Would they cease to express circatidal rhythms under such circumstances and, if so, would they retain the ability to recover their tidal memories, quickly or otherwise? It is practicable to address these questions in relation to the shore crab *Carcinus maenas* which is fairly ubiquitous in its geographical distribution and is indeed occasionally found living in non-tidal conditions. For example, as referred to earlier, it can often be found in non-tidal harbours that have only brief and intermittent connections with the sea, as ships enter or leave through a protective lock gate system. By collecting crabs from such a harbour system in South Wales, and recording their walking movements in constant conditions in the laboratory, it was found that they were always at their most active during the 'expected' night, their bursts of walking activity occurring progressively later during successive 24-h recording sessions (see Figure 3.1) (Naylor, 1960). So, isolated in the harbour, the crabs appeared to have lost the ability to express circatidal rhythmicity, which is endogenous in their conspecifics that are resident on the tidal shores outside the harbour gates. However, harbour crabs are able to reinstate their circatidal rhythmicity after a few days of exposure to tidal conditions on the shore (Figure 3.1) and, impressively, it was found (Naylor, 1960) that a circatidal rhythm could be quickly re-established in such crabs in response to no more than a single short, sharp temperature shock. It was known from other experiments with *Carcinus* from the open coast (see Naylor, 1963) that exposure to low temperature for a few hours was sufficient to reset the tidal clocks of crabs that had lost the ability to express tidal rhythms in the laboratory. Therefore it seemed reasonable to repeat the same experiment with harbour crabs, which showed that if they were cooled for a few hours and then monitored at room temperature afterwards, they promptly exhibited a rhythm of circatidal periodicity, exactly as was found to be the case in artificially reared crabs (Figure 2.5) (Naylor, 1960). Clearly the crabs had retained the ability to express a circatidal rhythm of behaviour, the phase of which was set by the time that they were removed from their low temperature experience. The conclusion, therefore, was that in the absence of tides the harbour crabs did not express tidal rhythmicity but they clearly retained the ability to do so. Evidence suggested that they would quickly be able to revert to normal rhythmic behaviour patterns if they found themselves outside the harbour gates in tidal conditions once more, as was found to be the case (Figure 3.1). Indeed, it was found that the same phenomenon occurred in winter when large crabs tend to migrate below tidemarks, and few are found between tidemarks at low tide (Naylor, 1962). Large *Carcinus* collected from below tidemarks in winter express only circadian rhythms of walking behaviour in constant conditions in the laboratory. Such crabs do experience changes in hydrostatic pressure associated with tidal rise and fall but, because of the low winter temperatures and lack of exposure to air at low tide, their rhythmicity is essentially circadian. They do, however, quickly resynchronize their circatidal clocks in spring when they move back into the intertidal zone and are exposed to the full range of tidal variables.

Such studies then prompted the question as to what happens in crabs that have lived in non-tidal conditions over much longer, evolutionary timescales. It was possible to address this question by carrying out experiments on a very close relative of *Carcinus maenas* resident in the Mediterranean Sea. Once thought to belong to the same species as *Carcinus maenas* that occurs commonly on open Atlantic shores of Europe, the species that lives in the non-tidal Mediterranean Sea was, in the mid-twentieth century, first designated separately as *Carcinus mediterraneus* and sometimes as *Carcinus aestuarii*. When recorded in constant conditions at the Stazione Zoologica in Naples, *Carcinus mediterraneus* showed only a circadian rhythm of walking behaviour, with most activity occurring during the times of 'expected' night (Naylor, 1961). The crabs showed no evidence of behaviour that could be construed as a legacy of tidally related activity. Moreover, whereas cooling of *Carcinus maenas* from a south Wales non-tidal harbour restarted their circatidal clocks, this was not the case in *Carcinus mediterraneus* from the virtually tideless Mediterranean Sea. Evidently, *Carcinus* that have lived in the Mediterranean for a sufficiently long period of geological time to have evolved as a separate species have, by the same process, also lost their tidal memories. Also, though fishes of tidal shores show locomotor rhythms of circatidal periodicity (Gibson, 1965), related Mediterranean species exhibit circadian rather than circatidal rhythmicity (Gibson, 1969).

*   *   *   *

In Chapter 2 it was shown that a wide range of animals from tidal shores exhibit tidal memories when isolated from their native beach, expressed as circatidal and circadian rhythms of locomotor activity and other behavioural and physiological processes. Then, in this chapter, a number of environmental variables associated with tidal rise and fall and day/night cycles have been shown to entrain those rhythms (see also Dunlap *et al.*, 2004). The biochemical and molecular processes involved in coupling zeitgeber action with clock function in marine organisms are largely unknown, but some parallels with what is known of circadian entrainment processes in terrestrial organisms will be discussed in Chapter 10. Here it can be conclusively stated that behavioural and physiological circatidal rhythms in coastal organisms, like their circadian rhythms, are driven by internal clockwork that is cued by external environmental cycles, which, like the Liverpool One O'Clock Gun, clearly serve as clock adjusters. In dense populations of animals there may also be an element of mutual entrainment of biological clocks (Harker, 1964; Bregazzi and Naylor, 1972; Broda *et al.*, 1985), animals that become spontaneously active at the most advantageous time stimulating into activity individuals that have poorer time-keeping ability. This mutualism improves the time-keeping ability of the population as a whole, provided there is also continuous input of time cues from the environment. The point is nicely illustrated by a story relating to timing of the One O'Clock Gun, not in Liverpool, but in Edinburgh before that gun was linked to the local astronomical observatory. It is said that the gunner would watch through a telescope the time of day on a clock in a shop window on Princes Street and at one o'clock the gun was fired. Immediately the gun was fired the shop-keeper was one of several who then emerged to check that their clocks were set correctly against the timing of the gun! Whilst mutual entrainment of this kind ensures synchrony of behaviour it does not necessarily ensure accurate phasing of synchronous timing, emphasizing the need for linkage to an observatory to provide a time-cue based on independent astronomical events. But, apart from the role of cyclical environmental variables as zeitgebers of biological rhythms there are other interactions between the two in relation to the navigational abilities of animals, phenomena to be discussed in the next chapter.

4

Clocks and compasses

One of the most spectacular aspects of biological clocks is their participation in celestial orientation.

Klaus Hoffmann, 1982

Numerous accounts of impressive navigational skills and homing behaviour shown by a wide range of animal species prompted Conway Morris (2003) to comment that 'despite our admiration, wonder, and – if we are candid – even awe, we can surely offer the following para-phrase; evolution happens'. What, beyond anecdote, is the evidence for such phenomena that generated this comment by an eminent evolution-ary biologist? Migratory birds are well known to navigate over long dis-tances between feeding grounds and breeding sites, often over thousands of miles, some even homing on small, isolated islands after extensive trans-ocean flights. Moreover, in many cases of such spectacular homing behaviour, one does not have to rely on anecdotal accounts, as Matthews (1968) demonstrated in early studies of the Manx shearwater (*Puffinus puffinus*). From a colony of shearwaters nesting on the island of Skokholm off the south west coast of Wales, individual birds were transported away in various compass directions, including westwards across the Atlantic Ocean, and released. The birds were observed as they were released and many quickly showed significant orientation towards home, apparently based on the environmental features around them, even in an unfamiliar location. Most reached their nest sites on Skokholm far more rapidly than they could have done by random searching. Indeed, one bird released on the east coast of North America reached Skokholm before the airmail letter sent to indicate the location and time of its release (Matthews, 1968). More recently, too, highly sophisticated experiments have been undertaken using satellite tracking to monitor the homing behaviour of oceanic birds to which miniature radio transponders have been attached. One trackway of the wandering albatross *Diomeda exulans* plots a 13-day flight over nearly 5000 km of ocean (Prince *et al.*, 1992). From its nesting site in South Georgia it flew north west to the Falkland Islands, then south west to Tierra del Fuego and returned more or less directly north eastward back to South Georgia. These forms of homing behaviour are natural phenomena that resemble human navigation on land, by sea or in the air, but without the use of human gadgetry. Shearwaters, and homing pigeons too, when taken to a locality with which they are unfamiliar, seem able to determine their correct homing direction quite quickly, with reference only to cues available to them in the novel locality. One possible explanation for this is that they are able to make a comparison between astronomical information, such as the position of the sun, moon or stars, at the new locality compared with home at a particular time of day. To make such a comparison they would require a biological clock that signalled to them where, say, the sun should be at a particular time of day at home, compared with its observed position and apparent orbit in the new locality. Then, having established their position in relation to home they would also require some form of compass mechanism to permit them to establish the correct bearing to return. Hypotheses to that effect have been investigated and some experiments relevant to those studies will be discussed later.

An alternative hypothesis is that birds carry with them and are able to read some form of map. In that event they would still require a compass to navigate home, but they might not need a clock. One line of research along these lines postulates that oceanic birds have an awareness of the pattern of the earth's magnetic field, a map that they carry and use to determine their homing directions. To test the hypothesis, experiments have been carried out on a number of oceanic bird species by attaching magnets to their bodies on the assumption that they would then be confused if they normally navigate in response to the earth's geomagnetic field. Albatrosses (Bonadonna *et al.*, 2003a, 2005), shearwaters (Massa *et al.*, 1991) and petrels (Benhamou *et al.*, 2003) have all been tested in this way and all navigated over very long distances to their nesting sites as normal, as if their navigation system was unaffected by the presence of an artificial magnetic field around them. However, notwithstanding the consistency of evidence so far against the hypothesis, the jury is still out regarding whether or not magnetic cues are involved in global navigation by oceanic birds (Papi, 2006).

A search is also underway to determine whether some birds might also navigate by a form of olfactory map (Papi, 2006). Particular success in this approach has been achieved in studies of Leach's petrel (*Ocean-odroma leucorhoa*) which for some years has been known to navigate locally back to its nesting burrow using its sense of smell (Grubb, 1974). More recently, too, the same species of bird has been shown to be able detect the smell of the volatile compound dimethyl sulphide which is produced by many phytoplankton species that form the basis of marine plankton food chains. Productive areas of the ocean surface which are rich in phytoplankton and the planktonic animals that feed thereon are favourite feeding grounds for Leach's petrels, attracted by the smell of dimethyl sulphide emitted abundantly there. Evidence that this form of navigation by the petrels has been perfected during evolution comes from experiments in which the sense of smell of a few birds was artificially and temporarily impaired (Bonadonna *et al.*, 2003b). Affected birds were unable to show homing behaviour when released over a feeding area. So, the apparently featureless surface of the ocean may present a varied olfactory landscape to marine birds with a keen sense of smell.

Many of the large, fully aquatic marine animals also make extensive migrations between feeding grounds and localities where they breed, including fish, turtles, seals and whales. Particularly striking is the migration behaviour of marine turtles (Carr, 1963; Hays *et al.*, 2003), notably the green sea turtle *Chelonia mydas*, one population of which nests on Ascension Island in the mid-Atlantic, just south of the equator. Out of the breeding season the turtles are to be found feeding on coastal vegetation of warm shallow seas along the coasts of Brazil, where individuals may remain for several years between breeding migrations. However, when they do migrate to lay eggs, adult female green turtles swim for a distance of about 2000 km into the Atlantic Ocean, eastwards towards their nest sites on the beaches of Ascension Island, before returning again to the coast of Brazil.

With a journey of 2000 km and a target island only about 8 km in length female green turtles achieve an amazing feat of navigation. As with marine birds and some fish such as the sockeye salmon (*Oncorhynchus nerka*) (Quinn and Brannon, 1982), it has been hypothesized that the turtles navigate during their trans-oceanic journeys by responding to the pattern of the earth's magnetic field. Are they in some way reading a geomagnetic map and fixing their position accordingly? In laboratory experiments there is evidence that this is the case, as was found by exposing sea turtles to artificial magnetic fields that simulated natural geomagnetic fields at various geographical sites. Orientated movements of the turtles under such conditions were consistent with the hypothesis that they were capable of magnetic navigation (Lohmann *et al.*, 2004;

Lohmann and Lohmann, 2006). Unfortunately, however, other experiments carried out on turtles in their natural environment raised questions about that interpretation. Green turtles with magnets attached to their bodies did not become disorientated in the earth's magnetic field and navigated as successfully as they normally do from Brazil to Ascension Island (Papi *et al.*, 2000). Furthermore, similar experiments have been carried out using Indian Ocean green turtles, with the same result. Turtles were found to navigate equally successfully to their nest sites on Mayotte Island, between the northern tip of Madagascar and the African mainland, whether or not they had magnets attached to their bodies (Papi, 2006). Like the oceanic birds that have been tested in the same way, therefore, trans-ocean navigational ability is unaffected by an artificial magnetic field around them. So, either they are not navigating in response to the earth's magnetic field or their responses to it override the local effects of an attached magnet. Certainly, recent evidence has also been presented which suggests that Caribbean spiny lobsters, *Panulirus argus*, are able to navigate in response to the earth's magnetic field (Lohmann and Lohmann, 2006). Earlier studies using tagged lobsters had shown that they forage at night, ranging several kilometres away from their home crevices where they hide by day. To test the hypothesis that the lobsters derive positional information in the earth's magnetic field, captive lobsters were exposed to artificial magnetic fields mimicking those at specific locations within their geographical range. Lobsters tested in a magnetic field replicating that north of their capture site moved south, but they moved north when placed in a field simulating that to the south of their location of capture. It is remarkable that lobsters are able to detect differences in the earth's magnetic field over such relatively short distances but, in these and other marine animals, more experiments will no doubt be soon forthcoming, seeking to resolve the conundrum of how general the phenomenon of animal navigation in a geomagnetic field might be. Importantly too, as will become apparent later, it will be necessary to establish the time base of navigational responses since, in a number of species, orientation has been found to vary over the lunar cycle (Morgan, 2001). For example, Lohmann and Willows (1987) showed that the sea slug *Tritonia diomeda* orientated precisely in the geomagnetic field during times of full moon but was disorientated during new moon periods. Moreover, when the magnetic field was reversed using an induction coil, the direction of orientation at full moon was also reversed. The moon evidently acted as a cue for directional orientation, perhaps facilitating migration of

Tritonia onshore during the reproductive period. Such findings prompt a return to the original question posed at the beginning of this chapter concerning position-finding behaviour in animals that might involve the use of a compass and a biological clock.

*   *   *   *

Early in the twentieth century entomologists marvelled at the precise sense of direction in ants as they explored away from and then returned to their nests, one explanation of the phenomenon proposing that ants possessed a mystical sense of direction. However unscientific that explanation now appears, it is still the line of least resistance of some non-biologists to account for similar phenomena today. A perhaps equally improbable early explanation suggested that the insects possessed a very exact kinaesthetic memory which implied that they recorded, evaluated and remembered every limb movement associated with directional changes during an outward journey, enabling them the make the correct reverse movements to return to their home nest. In fact, in the early twentieth century, it was demonstrated by very simple but ingenious experiments that ants use the position of the sun to enable them to navigate (see Hoffman, 1982). Ants on their way back to the nest were first shielded from the direct light of the sun and then, by the use of a mirror, were exposed to reflected sunlight, apparently coming from another direction in the sky. Under these conditions the ants adjusted their 'homing' direction predictably according to the apparent position of the sun, immediately making a course correction again when the mirror was removed. Several such experiments convincingly showed that the sun was the orientating stimulus, leading to the obvious suggestion that on short journeys away from the nest ants remembered the position of the sun and reversed their angle of orientation to it on the return journey. However, on longer journeys away from the nest the mechanism would not work, because the apparent position of the sun would change and the reverse angle of orientation would take the ants in the wrong direction. Because of the earth's rotation the sun appears to move across the sky at the rate of some fifteen degrees of arc every hour, so, if a journey away from the nest took several hours, reliance on a fixed angle to the sun would be misleading. On longer journeys correct orientation by this means could be achieved only if the ants made allowance for the passage of time, thus recognizing the apparent movement of the sun and adjusting their orientation angle accordingly. In short, compass navigation by the sun or indeed by any other celestial object can only be achieved by the possession of a clock, biological or otherwise. Only by the acquisition of a sense of time does orientation, the ability to follow a bearing, become true navigation which involves setting, adjusting and following a bearing.

If, initially, it was thought to be remarkable that animals might possess an internal biological clock to mark off intervals of time, the possibility that some might have a continuously consulted biological clock was deemed unlikely. It was not until the mid-twentieth century that evidence in favour of the idea came forward when time-compensated sun-compass orientation was demonstrated unequivocally in birds, including some marine species, many of which exhibit precise homing behaviour which appears to be clock-based (Kramer, 1952; Matthews, 1968; Schmidt-Koenig, 1975; Hoffman, 1982). Even when kept in outside cages at the time of year when they are expected to migrate, many birds show 'migratory restlessness' or 'intention movements'. At such times they hop in the direction of their normal migratory route, a form of behaviour that was found to be particularly convenient for study in starlings, which migrate annually between breeding grounds in north west Europe and wintering localities in England and Ireland, crossing the Baltic, North Sea and possibly the Irish Sea as they do so. Quite early in such studies, Kramer (1952) captured starlings that were about to migrate and restricted them to small circular cages in which the sight of landmarks was excluded. Their intention movements were then observed and found always to be in their migratory direction whatever the position of the sun throughout the day. The birds' orientation changed only when the sun's apparent position was changed by mirrors. The sun was not only the guiding cue, but the starlings were able to compensate for the movement of the sun evidently by the use of an internal clock mechanism.

The next step was to determine whether the timing mechanism was a simple hour-glass system or whether it was a true endogenous circadian oscillator. Was it a countdown timer set in motion by light changes at dawn or dusk, or was it a continuously consulted biological clock? To test which hypothesis was correct, birds were exposed to artificial day/night cycles with the onset of daylight advanced by 6 h and the migration direction tested daily. Under these conditions the birds gradually changed their migratory direction, taking several days to reach the 'correct' direction. If their clock mechanism was based on the egg timer principle they would have adjusted their orientation immediately. The fact that it took several days for the adjustment to take place suggests that their clock mechanism is a true circadian oscillator. Indeed, when starlings were kept in continuous light for several days, their circadian clock system free-ran slightly faster than the solar day, after which they misjudged the position of the sun and sought to head off in a predictably different migratory direction (Kramer, 1952).

The full story of how birds navigate remains to be elucidated, particularly concerning the demonstrably precise homing behaviour of species that travel over hundreds of kilometres of unknown territory. Not only is some aspect of navigation by the sun involved but, as we have seen, there is evidence too that some birds may use olfactory and possibly even geomagnetic maps to find their way about. The last two compass mechanisms need no time-compensation, but celestial navigation does, and it is this combination of clock and compass system that will now be explored in animals of the sea coast.

*   *   *   *

Mobile creatures of marine shorelines are much more amenable than birds or turtles to experimental investigation of the clock/compass problem. Many are widely dispersed at one stage of the daily or tidal cycles and are distinctly zoned at other times, requiring them to exhibit repeated homing behaviour that returns them to their preferred zone on the shore. Early indications of directional movements linked to a time sense in a coastal animal came from studies in the 1950s of the sand hopper *Talitrus saltator*, an amphipod crustacean, in one of its localities on beaches in Italy. By day the hoppers are normally found burrowed a few centimetres below the sand surface just above sea level. They emerge at night to forage and feed in dune systems away from the sea, which is virtually tideless along the Italian coastline. Investigations by Papi and Pardi (1953) began by asking the question as to how the hoppers returned to their preferred burrowing zone near the water's edge, thus avoiding the heat and desiccation of the midday sun (see also Pardi and Ercolini, 1986; Papi, 2006). If they were dug up in the middle of the day and placed on the open beach, or even in the sea, they quickly returned to the level of the beach where they preferred to burrow. Then, in laboratory experiments, Papi and Pardi (1953) released hoppers at the centre of circular arenas, the walls of which shielded the hoppers from all visual cues except for the sun and sky above. They found that the hoppers released at the centre of an arena moved immediately towards the circular wall in a direction that, in nature, would have taken them to the position of their burrowing locality on the beach. Even in such artificial surroundings they headed in the correct 'homing' direction, irrespective of the time of day. It seemed that they were navigating according to the position of the sun and were compensating for the apparent movement of the sun as they did so, as experiments using reflected sunlight confirmed (Figure 4.1). Papi and Pardi then carried out a critical experiment by testing the navigational

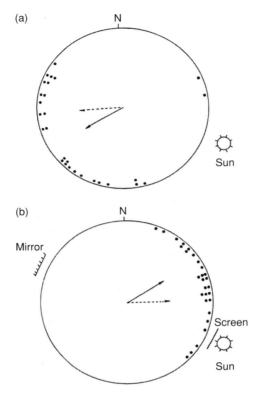

Figure 4.1 Sun-compass orientated escape directions (dots) of Mediterranean *Talitrus saltator* in an experimental arena from which the sun was: (a) visible directly; (b) reflected through 180 degrees. Solid arrows, observed mean escape directions; dotted arrows, expected mean escape directions (after Pardi and Ercolini, 1986; by permission of Taylor & Francis Ltd.).

ability of hoppers which they transported from the west coast of Italy to the east. In the new locality they continued to move in the compass direction appropriate for regaining the burrowing zone on their home beach, despite the fact that this took them inland to unfavourably hot and dry locations and not to moist sand at the sea edge. Clearly the hoppers were not navigating by local visual or other clues that would have guided them to a more favourable location. Thus there seemed little doubt that during daytime, when the experiments were carried out, Mediterranean *Talitrus* exhibited sun-compass homing behaviour and were able to compensate for the apparent movement of the sun by the possession of a biological clock mechanism of circadian periodicity. Moreover, the homing direction exhibited by *Talitrus* from any one beach appears to be partially inherited, since first and second generation laboratory-reared offspring exhibited approximately the same spontaneous escape directions as the parent stock when tested in a circular arena (Figure 4.2) (Scapini and Buiatti, 1985). During night-time, when *Talitrus* is in any event normally most active on the sand surface, the hoppers were also shown to navigate using a moon-compass, again on non-tidal Mediterranean beaches, initially by Papi and Pardi (1953) and Pardi and Papi (1953), and more recently by a number of other workers (see Meschini *et al.*, 2008) as will be discussed later. But *Talitrus saltator* occurs much more widely throughout Europe than in the Mediterranean and, in different localities, particularly regions subject to tides, it might be expected to behave differently. It also occurs in open Atlantic beaches where extensive stretches of moist sand are exposed at low tide, few more so than on the shores of the British Isles where, depending upon locality, tidal ebbs of up to 10 m vertical fall may occur at spring tides, revealing vast stretches of sand at low tide.

It was on such markedly tidal shores in north east England that Williamson (1951), around the time that Papi and Pardi began their studies in Italy, noted that British *Talitrus saltator* did not move away from the sea at night as they do in the Mediterranean, but foraged downshore towards the sea. On tidal shores it was found that displaced hoppers in such localities found their way back to their daytime burrowing zone near high water mark using different orientation cues from those used by conspecifics on Italian beaches. In contrast to the time-compensated responses to the position of the sun shown by Mediterranean *Talitrus*, British sand hoppers were found to orientate much more simply. At dawn they moved towards the dune/sky boundary, which they could see at the top of the beach. When they reached moist sand near the strandline, as distinct from the wet sand of the open beach, they then promptly burrowed. They did so by hopping up or down slight undulations in the beach and were therefore not responding to the general upwards slope of the sand surface. In the laboratory, hoppers were also shown to move towards an artificial dune/sky boundary, irrespective of the slope of the substratum that they were provided with. Such an orientation mechanism would, of course, return the animals to the strand line at dawn whatever the aspect of the beach upon which they found themselves. Moreover it would not necessarily require the use of a biological clock as a navigational aid. Sand hoppers on tidal shores in Britain therefore seemed to orientate more simply back to their preferred zone on the beach, quite differently from the navigational behaviour of the same species living on Mediterranean beaches.

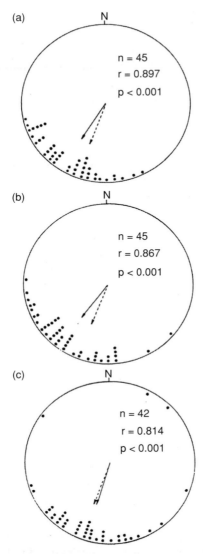

Figure 4.2 Inheritance of orientated escape directions (dots) of
Mediterranean *Talitrus saltator* tested in an arena from which the sun was
visible: (a) freshly collected adults; (b) first generation laboratory-reared
specimens; (c) second generation laboratory-reared specimens. Solid
arrows, observed mean escape directions; dotted arrows, expected mean
escape directions; n, number of animals; r, mean vector length (adapted
from Scapini, 1986, in Pardi and Ercolini, 1986; by permission of Taylor &
Francis Ltd.).

However, that is not to say that British *Talitrus saltator* do not possess a biological clock. Following up Williamson's (1951) finding that British hoppers were nocturnally active, it was later established that they clearly possess a circadian biological clock that controls the timing of their downshore foraging excursions (Bregazzi and Naylor, 1972). In the laboratory the locomotor activity rhythm of the animals was shown to continue for many days in constant conditions, delaying by only a few minutes each day (see Figure 3.4). The rhythm is driven by a true circadian oscillator system that is reset daily by the onset of daylight at dawn (Williams, 1980). The question then arose as to whether the biological timing mechanism of British *Talitrus saltator* was only used as an alarm clock which times the periods of locomotor activity or whether it too might have a wider application during the orientation process. Since Williamson (1951) had carried out his experiments during daytime when the amphipods would not normally be found on the sand surface and when it would be to their considerable advantage to head upshore and burrow, away from the attentions of shore-feeding birds, it seemed reasonable to ask how the hoppers would react to the shape of the dune/sky boundary at night, albeit by the light of the moon. At night, when on tidal beaches they normally tend to move downshore to forage for food, do the hoppers switch off or reverse their orientation response to the dune/sky boundary? Answering this question entailed carrying out experiments in which hourly observations were made continuously throughout the day and night (Edwards and Naylor, 1987). By this means, around-the-clock observations of the orientation responses of sand hoppers were carried out, testing their reaction to an artificial dune/sky boundary of the kind used by Williamson.

During the early part of the day the hoppers did indeed move towards the dune/sky boundary, as Williamson had shown. However, from the late afternoon until just before dawn they were completely indifferent to the boundary (Figure 4.3). The pattern of change of response occurred not only when the hoppers were kept in a normal cycle of light and dark, but also when they were kept in continuous dim light. The pattern of changing responses therefore appeared to be under biological clock control, suggesting that a circadian clock mechanism of *Talitrus saltator* is used to switch on the upshore navigation response just before dawn and to switch it off again before nightfall (Figure 4.4). At this stage in the experiments then, it was clear that the homing mechanism of hoppers on tidal beaches was under the combined control of orientation towards the dune/sky boundary and a biological clock but it was not known whether the locomotor activity rhythm and the

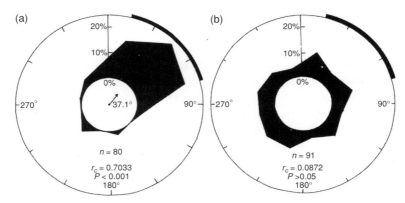

Figure 4.3 (a) Orientated escape directions in an experimental arena of *Talitrus saltator* from a tidal beach in Britain, tested at 12.00 h. A total of 80 amphipods were tested individually, with a vertical 60 degree arc of black cardboard centred at 45 degrees on the perimeter wall of the circular arena. The percentage of animals which reached each 30 degree arc of the perimeter wall after release at the centre is shown along the radius. The arrow at 37.1 degrees is the observed mean escape direction, the expected mean escape direction being 45 degrees. The value of r indicates highly significant non-uniformity of the circular distribution of the escape directions by the Rayleigh test; (b) random escape direction shown by 91 amphipods tested at 04.00 h (after Edwards and Naylor, 1987).

orientation rhythm were controlled by the same or by different daily clocks. Also it was necessary to find out how the hoppers made their way down the beach to forage after their alarm clocks had awakened them to emerge after nightfall, particularly since they ignored the dune/sky boundary at that critical time.

In the round-the-clock experiments (Edwards and Naylor, 1987), it would have been a neat outcome if during the hours of darkness the sand hoppers actively moved away from the artificial dune/sky boundary. Then one could have formulated the simple hypothesis that a circadian clock of *Talitrus* simply switched on the orientation response by which hoppers were attracted to the dune/sky boundary around dawn and switched on a reversal of that response when the hoppers emerged at dusk. But since they were simply indifferent to the boundary at night, it raised further questions as to whether they moved downshore by random dispersal over wet sand or did so by a different, more specific orientation response. Addressing these questions, it was found that *Talitrus* on British tidal beaches did indeed use an additional type of response that permitted them to complete their daily cycle of behaviour (Mezzetti *et al.*, 1994). By day the hoppers on tidal beaches move away from a light source,

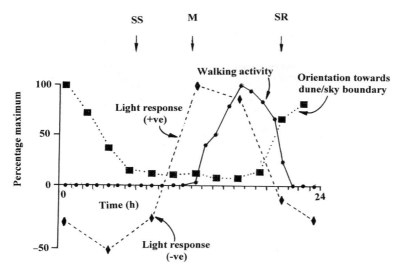

Figure 4.4 Clock-controlled locomotor and orientation rhythms of the sand hopper *Talitrus saltator* on a British tidal beach. When first active at night they move downshore, attracted towards light reflected off the sea. Around dawn they become less active and move upshore, attracted by the dark/light boundary of dunes against the sky. Solid line, locomotor activity; dashed line, response to light; dotted line, response to dune/sky boundary; SS, expected sunset; M, expected midnight; SR, expected sunrise. Values plotted as percentage maximum recorded during an expected 24 h cycle (after Naylor, 2002).

consistent with their inclination to burrow at that time. However, it was found that at night they were attracted to a light source, a response interpreted to mean that, as they emerged to forage down shore at night, they moved towards the brighter surfaces of the sea and wet sand and away from the less reflective dunes.

The nocturnal pattern of foraging behaviour in *Talitrus saltator*, whether on tidal Atlantic shores or non-tidal Mediterranean beaches, has presumably evolved as an adaptation that minimizes predation by daytime-feeding birds and the risk of desiccation on the sand surface by day. In these circumstances any mutation that results in behaviour patterns that help to ensure that the hoppers avoid unfavourable conditions would be selected for. Strictly speaking, acceptance of this argument requires experimental demonstration that survival of the hoppers in their natural environment is indeed enhanced when compared with populations that might be prevented from burrowing by day. Experimental tests of such hypotheses are difficult to carry out, but the circumstantial case is strong that over evolutionary time, *Talitrus saltator* has acquired an inherited biological clock capability that allows it to emerge on time and to switch on and off behavioural responses that ensure that it is in the right place at the right time. Even so, the amphipods evidently still retain evolutionary flexibility in their behavioural repertoire, since even on Mediterranean coasts they have been shown to utilize landscape features. The use of astronomical cues as originally described by Papi and Pardi (1953) appears to have been coupled with the use of landscape cues by *Talitrus saltator* particularly on stable beaches around the Mediterranean. Indeed, Ugolini *et al.*, (2005a) not only confirmed the influence of black and white artificial landscapes on zonal orientation in Mediterranean *Talitrus* but also showed that vision of a blue and green artificial landscape, representing the sea/land boundary, also affected orientation. Moreover, the response to a blue and green boundary overrode the response determined by the sun-compass mechanism, suggesting that colours of the natural environment greatly contribute to directional choice. In addition, it has been shown that hoppers on unstable beaches subjected to wind and wave erosion tended to rely solely on the use of landscape cues (Gambinieri and Scapini, 2008).

So, 50 years after the initial observations were made it became clear that, wherever it occurs, *Talitrus saltator* uses an internal clock as a basis for its ability to show homing behaviour. Directional movements to their burrowing zone have been shown to be in response to the position of the sun, to landscape features and, in a less well-understood manner, to the position of the moon (Ugolini *et al.*, 2005b; Meschini *et al.*, 2008). There seems little doubt therefore that *Talitrus* possesses circadian clock capability which times its solar day pattern of walking activity and its sun-orientation mechanism, and evidence is emerging that it may also possess circalunidian (approximately lunar day) and circalunar (approximately lunar month) clockwork as a necessary basis for its navigational ability in relation to the position of the moon (Meschini *et al.*, 2008). However, the question still remains as to whether the biological clock system involved in sun-related homing behaviour was the same, or different from, that which controls the day/night rhythm of walking activity. Early observations indicated that the two rhythms have certain features in common, since both are synchronized by light/dark changes and are similarly phase-shifted when exposed to a rise in temperature (Bregazzi and Naylor, 1972). Then some subsequent observations suggested that the locomotor activity rhythm of *Talitrus* may have a slightly different periodicity from that of the orientation rhythm (Edwards and Naylor, 1987). Clearly, it was necessary to record both rhythms simultaneously, using hoppers from the same population, and to investigate the separate responses of the two rhythms to various phase-shift treatments (Edwards and Naylor, 1987). Only then could valid conclusions be drawn on the question of separate or coupled rhythms of locomotion and orientation behaviour in the zonation behaviour of *Talitrus saltator* and such experiments were eventually carried out by Scapini *et al.* (2005). In those experiments a number of sand hoppers were first placed individually in separate containers, so that they did not influence each other, in constant conditions in the laboratory. Hoppers kept as a group in the same recording system in the laboratory tended to interact such that early risers awakened the late sleepers, leading to mutual synchronization of their circadian clocks. However, in isolation, their biological clocks slowly drifted out of phase from each other in a true, approximately daily, circadian manner, such that after two or three weeks rhythmic bursts of walking activity of individual hoppers occurred at any time throughout a 24-h interval (Scapini *et al.*, 2005). The orientation responses of the 'randomized' hoppers were then tested individually, the body clocks of many of them being seriously out of phase with real time. Would they still show 'homing' behaviour when released at the centre of an experimental arena in the laboratory and, if so, in which direction would they go? They did show 'homing' behaviour but in different directions that could be accounted for only by the disrupted timing of their body clocks. If a hopper was exposed to the sun during a real-time morning and its disturbed body clock signalled that it was really late afternoon, it 'assumed' that the sun was in its late afternoon position. Accordingly, even in the laboratory arena, it headed in a direction which would have taken it back to the sea's edge if it had been displaced inland during the late afternoon. These findings clearly suggested that the sun-based orientation rhythm and the daily walking rhythm of *Talitrus* are timed by the same circadian body clock, a conclusion also reached by Ugolini (2003) and Ugolini *et al.* (2007). In those studies a comparative study was made of the circadian locomotor activity rhythm and the 'chronometric' orientational mechanisms of compensation for the apparent movements of the sun and moon. It was found not only that the circadian locomotor activity rhythm and the solar-day compensation mechanisms appeared to be controlled by the same circadian oscillator, but that the lunar-day compensation mechanism appeared to act quite independently.

*   *   *   *

The evolutionary advantages of clock-based activity and navigation are particularly clearly seen in another species of sand hopper *Orchestoidea tuberculata* from sandy beaches of Chile. In laboratory studies it was shown that where they occur together adults of this amphipod crustacean compete strongly with juveniles by eating them. In such circumstances

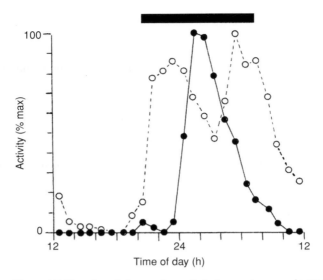

Figure 4.5 Mean hourly locomotor activity (percentage maximum) of the sand hopper *Orchestoidea tuberculata* over a standard 24 h interval, based on 4 days of recording in constant conditions in the laboratory of: adults, closed circles; and juveniles, open circles. Horizontal black bar indicates expected night (after Kennedy *et al.*, 2000; by permission of Springer Science and Business Media).

there would be clear advantages for the survivorship of the species as a whole if biological mechanisms were in place to ensure that, as much as possible, the two life stages were kept apart and this indeed was found to be the case (Kennedy *et al.*, 2000). Like their European relative *Talitrus saltator*, adult and juvenile *Orchestoidea tuberculata* exhibit daily patterns of burrowing and surface foraging, complemented by daily changes in response to landscape clues. Unlike their European relatives, however, adult and juvenile Chilean hoppers behave quite differently from each other. Adults emerge to forage in the middle of the night, whereas juveniles are active on the sand surface around dusk and dawn. Moreover they continue with this pattern of behaviour even in constant conditions in the laboratory (Figure 4.5). In addition, when they emerge the adults forage downshore, whilst juveniles do so upshore. Moreover, when they burrow, adults do so high on the shore, and the juveniles are found in a zone rather lower down the beach than the adults. In fact adults and juveniles of the Chilean sand hoppers behave almost as separate, closely related species, expressing complementary circadian rhythmic behaviour patterns that that appear to have been selected for and which, by and large, preclude them from directly competing with each other for space and food.

The sand hoppers described, whether from Europe or South America, are essentially semi-terrestrial animals that exploit the inter-tidal zone for food. Yet though they possess circadian, and possibly circalunidian and circalunar, timing systems that help navigate to their preferred burrowing zones on the beach, there is little evidence that they possess circatidal clocks that assist in their position maintenance behaviour. For good examples of animals that use tidal timekeeping to help maintain their preferred zones on the shore clocks one has to look elsewhere, to the truly marine species of the intertidal zone proper.

* * * *

Use of a face mask when swimming between tidemarks during high tide shows that many animals can be seen moving freely over the substratum or in the water column. Some of these species are normally not seen at low tide, when they hide or burrow, but they emerge in large numbers as the tide floods the shoreline. Indeed some migrate upshore with the rising tide, but most return to a preferred zone between or below tidemarks at low tide. For example, on an exposed sandy beach in South Africa McLachlan *et al.* (1979) reported that all epifaunal and shallow-burrowing inter-tidal macrofauna appear to undergo tidal migrations, the animals move upshore with the incoming tide and downshore during the tidal ebb. In molluscs such as *Donax sordidus* and *Bullia rhodostoma* the movements result solely from behavioural responses to changing physical conditions, but in mysid and isopod crustaceans tidal migrations were considered to be partly driven by endogenously controlled rhythmic behaviour. Certainly, on Californian beaches it was established by Enright (1963, 1975) that several crustacean species moved up and down the beach during a tidal cycle, partly driven by endogenous circatidal rhythmicity. Such biological clock-controlled rhythmicity was replicated in the laboratory in the amphipod *Synchelidium* sp., the isopod *Excirolana chiltoni* and the anomuran crab *Emerita analoga*. On European sandy shores, too, it is well established that the isopod crustacean *Eurydice pulchra* behaves in a similar manner, also under the control of circatidal biological clock-work (Jones and Naylor, 1970). In addition, the common green crab of many shores throughout the world, *Carcinus maenas*, is also such a tidal migrant. Small juveniles of this species tend to occur near high water mark on rocky shores, but larger juveniles and adults shelter beneath rocks at low tide and, during repeated observations by divers throughout rising and falling tides, can been seen to emerge and often follow the edge of the rising tide. As noted in Chapter 2, however, the crabs are less commonly seen at the water's edge at high tide and during the ebb (Figure 2.1); at those times they appear to seek shelter or move downshore in anticipation of the falling tide. Only during night-time are individuals seen, or heard, moving around the open shore at low tide. As discussed in Chapter 2, laboratory recordings of crabs in constant conditions show that their bursts of walking activity occur spontaneously at the times when they expect midnight and high tide, indicative of control by internal clocks of circadian and circatidal periodicities, the latter permitting anticipation of the tidal ebb. In the crabs, however, and some other crustaceans referred to above, as in sand hoppers, position maintenance on the shore is achieved, not simply by their cycle of activity and rest, but also by temporal changes in orientational behaviour. As examples among the Crustacea, the amphipod *Synchelidium* (Enright, 1963, 1975; Forward, 1980), the isopod *Eurydice* (Warman *et al.*, 1993b) and the crab *Carcinus* (Warman *et al.*, 1993a) all appear to exhibit endogenous rhythmic changes in their responsiveness to light as part of their behavioural repertoire involved in the avoidance of stranding at unfavourable tidal levels on the shore.

In laboratory experiments it was found that larger migratory crabs were attracted towards a light source at the time of the rising tide, but repelled by light at high tide and during the ebb (Warman *et al.*, 1993a). During the day such behaviour would bring hidden crabs into the open and could contribute to the maintenance of position of crabs in well-lit areas in shallow water at the edge of the flooding tide. Moreover, clock-controlled reversal of the light response whilst the crabs were still active at high tide and during the early ebb tide, would clearly promote shelter-seeking behaviour and downshore movement to deeper, less well-lit, coastal waters (Figure 4.6a). Again the process of evolution appears to have selected biological clock-controlled behaviour and orientation patterns that have a genetic basis, which help to maintain shore crabs in optimal conditions whatever the state of tide. Here though, the clock-controlled pattern of orientation to light is not switched on until the stage in the development of crabs when they begin to exhibit tidal migrations. Juvenile *Carcinus maenas* during their early development phase after settlement from the plankton spend all their time between tidemarks and at all times move away from the light, consistent with their habit of remaining hidden beneath stones at high levels on the shore (Warman *et al.*, 1993a).

Swimmers on European beaches may also encounter swarms of small isopod crustaceans, *Eurydice pulchra*, that normally feed on other

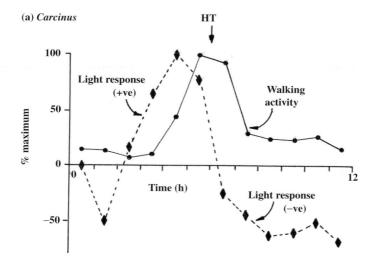

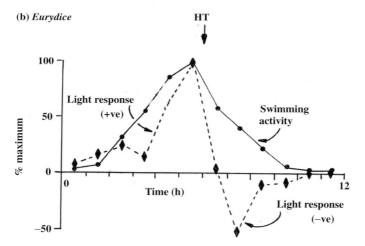

Figure 4.6 Clock-controlled orientation towards or away from light during high tide: (a) by walking *Carcinus maenas*; (b) by swimming *Eurydice pulchra*, when kept in constant conditions in the laboratory. At the end of their active sessions during times of expected high tide each species spontaneously moves away from light, as they would on the shore when hiding beneath boulders or in sand. Solid lines, locomotor activity; dashed lines, responses to light; HT, expected high tide. Values plotted as percentage maximum recorded during expected tidal cycle (redrawn from Warman *et al.*, 1993 a, b).

small animals in inshore plankton and which are not averse to biting human skin. At low tide, found buried in sand to a depth of a few centimetres in the upper reaches of the shore, the isopods emerge to feed in the water column during the flooding tide. After high tide they re-burrow in the sand in their preferred zone above mid-tide level, usually without being carried farther downshore. When kept in laboratory aquaria with sand and circulating seawater *Eurydice pulchra* shows a very clear circatidal pattern of behaviour, swimming for a while during expected high tide and burrowing at times of expected low tide (Figure 2.3). As was shown in Chapter 3, they are able to maintain such behaviour by possession of tidal time-keeping ability which can be reset by tidal (12.4-h) cycles of simulated tidal agitation. The tidal rhythm of swimming and burrowing helps partly to ensure that the isopods are in the right place at the right time, irrespective of tidal state, but the animals also exhibit a pattern of responsiveness to light that varies over the tidal cycle, which further enhances the zonal recovery process. When *Eurydice pulchra* first emerges from the sand during the flooding tide they are attracted to light at the sea surface to which they swim by positive phototaxis, placing them in wave induced onshore currents. In contrast, during the ebb tide, when their bout of swimming activity begins to subside, they are repelled by light, exhibiting negative phototaxis and swimming down into bottom currents which transport the isopods lower down the shore where they burrow in advance of being exposed at low tide (Figure 4.6b) (Warman *et al.*, 1993b). This interpretation of the rhythmic emergence and swimming behaviour of *Eurydice* is convincingly supported by the pattern of catches of the isopod taken at and near the water's edge during a cycle of tidal rise and fall (Warman *et al.*, 1991a). Throughout a tidal cycle the isopods were sampled in 20 cm of water at the tide edge, and at the surface and near the bottom in 1 m of water approximately 15–20 m offshore. Five samples were taken every hour at each depth, using a hand-net over a standard 50 m traverse parallel to the shoreline. The numbers of isopods taken at each sampling location were then considered in relation to the up- and downshore currents and the wave break point. Numbers caught were always greater in the surf zone at the water's edge than near the top or bottom offshore. In the surf, greater numbers were taken during spring high tides than during neaps, but in each case numbers of isopods caught increased up to the time of high tide, falling off more or less symmetrically after that time. Numbers caught were fewer between the wave break point and the tide edge but in that location greatest numbers occurred near the surface before the time of high tide, where the isopods exploit the wave-driven onshore current. In contrast, greatest numbers were taken near the bottom after high tide, where wave-driven downshore currents are exploited at that time.

For fully aquatic animals such as *Eurydice pulchra* that inhabit the upper reaches of a beach a further adaptive requirement for them is to avoid the risk of being stranded high and dry on the beach when the level reached by high tides falls from springs to neaps. 'Neaping' is a risk for intertidal migrants and *Eurydice pulchra* is an example of such an animal that has evolved adaptations to minimize that risk. At spring tides during the new and full moon periods the isopods can be found by digging and sieving sand when the tide is out at a level on the shore which is not normally reached by high tides at neaps, that is between high water neap (HWN) and high water spring (HWS). Some clearly re-burrow at that level after swimming during the high spring tides (Hastings, 1981b). So, if they do not emerge to feed during the next few days on subsequent high tides they may find that high tides do not cover them as spring tides reduce in height to low amplitude neaps. Under such circumstances the isopods would be left high and dry, and therefore at risk of death by desiccation, if not starvation. This takes us back to a question raised in Chapter 3 concerning the advantage to *Eurydice* in possessing a tidal clock. Initial findings suggested that they relied solely on immersion by the rising tide to time their emergence for high tide swimming, using their internal clocks only to return safely to the sand after swimming during high tide. But with the risk of being neaped, is it possible that they do use their biological clocks to time their emergence from the sand at certain times of the neaps/springs cycle?

To address this question a number of the isopods were maintained in an aquarium with sand and running seawater in constant conditions in the laboratory and their swimming activity was recorded continuously. The experiments continued for six weeks, with the animals receiving no cues as to the time of day or tide. Repeatedly, at and just after the days of the highest spring tides, swimming activity was spontaneously four times greater than during the remainder of each neaps/springs cycle (Alheit and Naylor, 1976). Such behaviour is clearly adaptive in ensuring that the bulk of the population swims during reducing spring tides and therefore moves downshore before the onset of neaps. Because peak high tide swimming behaviour occurs at and just after maximum spring high tides, which occur at approximately 14-day intervals, this aspect of position maintenance behaviour indicates that the isopods have not only circatidal and circadian biological clocks, but also clocks of approximately semilunar periodicity. Lunar periodicities in behaviour, whether directly moon-related or indirectly through the neaps/springs cycle of tides, will be discussed in greater detail in a later chapter. Suffice to say at this point that there is experimental evidence that the neaps/springs pattern of swimming by *Eurydice pulchra* is truly endogenous, driven by an internal biological clock system of circasemilunar periodicity (Reid and Naylor, 1985, 1986).

It is difficult to prove, but intuitively attractive, to suggest that endogenous locomotor activity rhythms and patterns of rhythmic responses to environmental variables of tidal, daily and semilunar periodicity seem to endow coastal animals with evolutionary advantages. Yet rigidly programmed behaviour patterns could be disadvantageous against a background of environmental variability that is sometimes unpredictable. Cloud cover can significantly change the times of dusk and dawn, and wind speed and direction can affect the timing and height of tides. Tidal currents may also vary unpredictably according to changes in the weather. How do intertidal animals cope with such unpredictable, or even catastrophic, events that might carry them offshore where they may be ill-adapted to survive? The endogenous free-running rhythms of the animals concerned, pre-fixed by 'circa', are by definition only approximate to the periodicity of the equivalent environmental cycle. Indeed they are repeatedly fine-tuned on each cycle by cues of environmental change, thus allowing some flexibility of response to minor environmental perturbations, such as sporadic changes in the timing of tides and gradual changes of dusk and dawn. However, occasional storms do sometimes displace intertidal species to unfavourable habitats above or below tidemarks. Intertidal species stranded high and dry above high water mark are not usually able to return to their preferred zone between tidemarks, but recovery from displacement below tidemarks can be achieved. For example, *Eurydice pulchra* is sometimes found below tidemarks accompanied by a related, deeper water species, *E. spinigera*, with which it does not normally compete. Does *E. pulchra* exhibit any kind of behaviour that might permit it to take avoiding action if it found itself accidentally displaced seawards and in competition with its subtidal neighbours? This question was addressed by maintaining mixed populations of *E. pulchra* and *E. spinigera* in the laboratory and exposing them to recordings of the sound of breaking waves. By this means the two species could be sorted quite simply, since *E. pulchra*, but not *E. spinigera*, were attracted towards the recorded sounds, a response that would in nature return them to their natural habitat between tidemarks (Jones and Hobbins, 1985). Similar responses to underwater sound have been observed in crab larvae returning from the open sea to settle on the shores of North Island, New Zealand (Jeffs *et al.*, 2003). Artificial sources of the natural underwater

Plate 1 The earth–moon system viewed from a spacecraft, with the moon closer than the earth to the camera. A NASA Galileo photo, from Head (2001), by permission of Springer Scientific and Business Media.

Plate 2 *Carcinus maenas*: A cosmopolitan intertidal crab which, in the laboratory, matching its behaviour on the shore, exhibits biological clock-controlled bouts of locomotor activity during expected high tide and night-time.

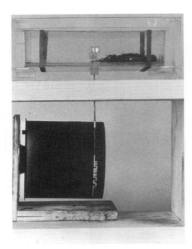

Plate 3 A simple 'rocker-box' actograph recording the circatidal pattern of locomotor activity of the common shore crab *Carcinus maenas* on a 24 h rotating smoked drum.

Plate 4 *Helice crassa*: A common burrowing mud crab in the high intertidal of soft shores in New Zealand. Often anecdotally reported to emerge as the tide falls, round-the-clock collection in pit-fall traps and laboratory recordings of behaviour show them to be most active during times of high tide, under the control of circatidal biological clocks, sometimes modulated by circadian periodicity.

Plate 5 *Macaca fascicularis*: The long-tailed macaque monkey digging for ghost crabs (*Ocypode* sp.) at low tide on a Malaysian beach. An occasional terrestrial predator in the intertidal zone. Photo by Stephen Hanbury.

Plate 6 *Talitrus saltator*: An amphipod crustacean of European sandy beaches which burrows in a narrow zone near the strandline by day, emerging to forage at night. Laboratory studies indicate that it uses internal circadian clockwork to time its pattern of emergence and changes in orientation behaviour throughout the 24-h period, which ensures its zonal recovery.

Plate 7 Actograph used for recording the circadian rhythm of walking behaviour of the sandhopper *Talitrus saltator*. Activity events are triggered by photosensitive devices when the hopper crosses a Perspex bridge between two containers of sand into which it burrows when inactive.

Plate 8 Divers recovering a resin cast of the burrow of a Norway lobster *Nephrops norvegicus*.

Plate 9 *Nephrops norvegicus*, the Norway lobster, emerging from its burrow in muddy sand.

Plate 10 *Eurydice pulchra*: A common sand-beach isopod crustacean of European shores which burrows at low tide, emerging to swim at high tide, particularly during spring tides. The swimming patterns are replicated in the laboratory, under the control of circatidal and circasemilunar biological rhythmic mechanisms.

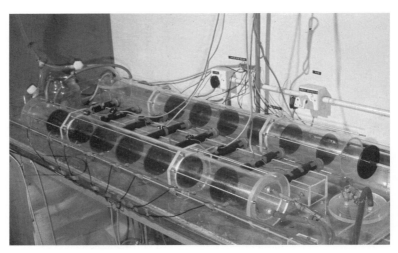

Plate 11 Apparatus used to record the locomotor activity patterns of crabs *Carcinus maenas* when exposed to cyclical changes of hydrostatic pressure. Movements of eight crabs in each pressure cylinder are monitored by photosensitive devices functioning across each compartment.

Plate 12 Apparatus attached on a rocky shore to record the tidal movements of the limpet *Patella vulgata*. A small circular magnet fixed at the apex of the limpet shell triggers a reed switch anchored permanently above the home scar. In the study locality limpets made feeding excursions at low tide and, when recorded in constant conditions in the laboratory, exhibited similar feeding excursions at the time of expected low tide.

Plate 13 *Sterna fuscata*: Sooty or Wideawake Terns nesting on Ascension Island in the mid-Atlantic, to which they return to breed usually every tenth full moon. Photo by John Hughes.

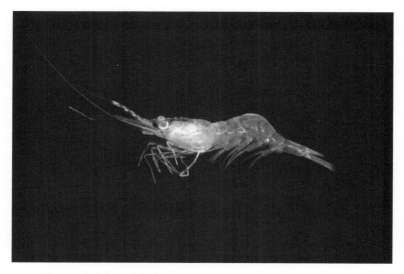

Plate 14 *Dichelopandalus bonnieri*: A vertically migrating prawn occurring off British south and west coasts which swims up off the seabed into the plankton at night, probably under circadian clock control.

Plate 15 Planktonic early and late zoea, and megalopa larvae of the shore crab *Carcinus maenas*. Early zoeae are released rhythmically and swim upwards rhythmically at times of high tide. The behaviours are replicated in the laboratory, indicating that they are under biological clock control. Photo by Douglas Wilson, from Hardy (1956), by permission of Mrs Hester Davenport.

Plate 16 *Uca pugilator*: A species of fiddler crab which is active at low tide on beaches of the east coast of USA and which, in the laboratory, exhibits a circatidal rhythm of locomotor activity with peaks at times of expected low tide. Larvae of several species of *Uca* are released during high spring tides, around times of full and new moon.

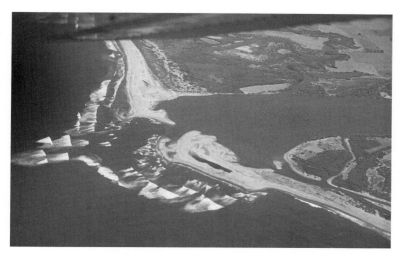

Plate 17 Aerial view of the mouth of the Rio Baluarte, Sinaloa, Mexico, with *esteros* meandering north and south from the main estuary. Penaeid prawn larvae travel from the open ocean into the estuary and *esteros*, en route to saline lagoons where they develop to pre-adult stages before making the reverse journey to the Pacific Ocean.

Plate 18 Low-level aerial view of a prawn fishing cooperative in western Mexico. The fishing trap (*tapo*), erected across the *estero*, catches pre-adult penaeid prawns when they attempt to make their way from the nursery site in a saline lagoon to the open Pacific Ocean beyond.

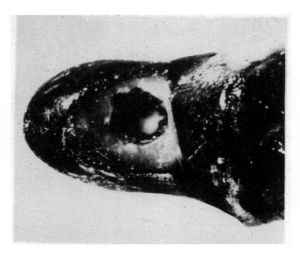

Plate 19 Dorsal aspect of the eyestalk of the crab *Carcinus maenas* with a circular hole cut into the cuticle to expose the white sinus gland in situ on the upper surface of the medulla interna of the optic tract.

sounds of wave action on coastal reefs were shown to attract newly settling crab larvae, particularly, it was found, at times of the moon's first
and third quarters. These were the times when neap tides began to rise at
nightfall, and the larvae swam upwards into the water column at those
times to be carried ashore by the flooding tides.

*   *   *   *

In many of the examples already referred to in this chapter the homing
behaviour described is achieved by 'zonal recovery mechanisms' (Pardi
and Ercolini, 1986), whereby the animals concerned are able to move
away from and return to their preferred zone between tidemarks dependent upon the precise state of tide. But some coastal animals such as
Patella vulgata, a common limpet of European shores, possess even more
precise homing behaviours that again rely on their biological clocks to
ensure being in the right place at the right time (Plate 12). Attempts to
dislodge large limpets from a rock surface quickly reveal that they are
often seemingly embedded in a rock depression that matches the shell
outline. In fact they are seated in their 'home scar', to which the limpet
returns after feeding excursions, abrading the rock surface slightly with
the edge of its shell each time it attaches itself tightly with its adhesive,
muscular foot. Feeding occurs at particular states of tide or times of day,
during night-time low tides, daytime low tides or at high tide, depending
upon geographical locality. They move freely, grazing on films of algal
cells on rock surface, secreting mucous trails as they do so, trails that
they use to navigate back to their home scars when they have finished
feeding. In response to speculation that such homing behaviour might
be at least partly under the control of a limpet biological clock *P. vulgata*
attached to small boulders were studied in the laboratory in continuous darkness, high humidity and away from the influence of tides (Della
Santina and Naylor, 1993). In the recording chambers a small magnet
was attached to the apex of each limpet and an electromagnetic switch
was fixed to overhang the home scar. With every move away from or back
to its home scar, each limpet under investigation was made to signal to
a computer a continuous record of its movements. As on the shore from
which they were collected, the animals in the laboratory showed greatest
activity away from their home scars at times when they expected low
tide, particularly at night (Figure 4.7), indicative of a tidal and probably
a daily time sense too. Evidently precise homing behaviour is of sufficiently valuable adaptive advantage that, over evolutionary timescales,
even slow-moving, tough-shelled limpets, like many highly mobile, more
fragile crustaceans, have acquired biological clock systems that seem to
enhance their chances of survival.

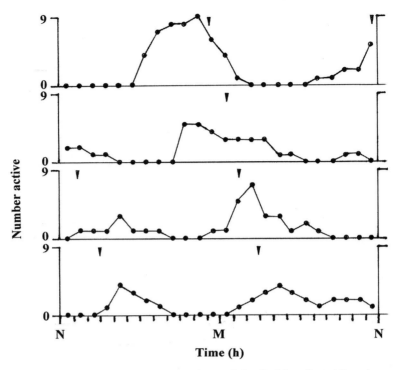

Figure 4.7 Greatest numbers per hour of nine freshly collected limpets *Patella vulgata* away from their home scars over four days in constant conditions in the laboratory. Arrows, times of expected low tides; M, expected midnight; N, expected noon (redrawn from Della Santina and Naylor, 1993).

It is relatively easy to understand how behavioural adaptations in coastal animals have arisen in response to the twice-daily rise and fall of tides. However, reference has also been made to the intriguing possibility of adaptation to the less dramatic differences that are apparent on a fortnightly cycle between neap and spring tides, further consideration of which will be undertaken in the next chapter.

5

Lunar and semilunar biorhythms

. . . our natural satellite regulates the life (on earth), through the stability of the terrestrial rotation axis, through the tides, and perhaps through other subtler effects that still need to be better understood.

Cesare Barbieri, 2001

Life on earth probably dates back at least 3500 million years. That would have been about 1100 million years after the earth was formed at the time of the origin of our solar system, thought to be about 4600 million years ago, another 5000 million years or more after the 'Big Bang' singularity at the beginning of the universe (Hawking, 1988). As the original earth cooled and its rock crust began to form, the moon was created by a process now widely accepted as the 'Big Splash' (Benn, 2001). Over 4000 million years ago, some say as little as 30 million years after the earth was formed, it is thought that the proto-earth was impacted by a cosmic object called a planetesimal of about the size of the planet Mars. Impacts of cosmic matter on the earth's surface, in the form of asteroids, comets and meteorites, have occurred many times in geological history, but none could have had such an impact on earth as a lump of matter at least one tenth of the size of earth itself.

Much later another highly significant impact event occurred at the end of the Cretaceous Period, a mere 65 million years ago. Rocks laid down at that time contain a very thin layer of the element iridium which is common in meteorites, but otherwise rare on earth. A meteorite colliding with earth at that time is thought to have resulted in a dark cloud of dust, rich in iridium, which covered the whole planet for several months until the dust settled. During that period it is estimated that daytime light intensities would have resembled those on moonless nights, resulting in the demise of many plant species that require light for photosynthesis, and the consequent death of many groups of animals that depended upon the plants for food and the animal carnivores that depended upon them. The dinosaurs died out during that time, an event often said to be directly related to the meteorite impact, their disappearance perhaps further hastened by a dramatic cooling of the earth's atmosphere that would have been generated by the dust cloud, equivalent to the effects of a so-called 'nuclear winter'. There is no doubt that massive effects were felt on Earth from the impact of a celestial body, which may have been no more than 20 km in diameter. On that basis, the impact of a Mars-sized object nearly four billion years earlier would have been colossal. Indeed, it is considered that a grazing collision of the planetesimal splashed into orbit around the earth a huge mass of debris from the earth's mantle that coalesced to form nothing less than the moon itself. On this, or any other explanation of its origin, the moon pre-dates life on earth by several hundred million years, and therefore either directly by its reflection of sunlight, or indirectly through its gravitational influence on ocean tides, it can be expected to have influenced living organisms throughout their evolution. More than this, however, the moon may have had a significant role in the origin of life itself.

How life began has long been a matter of speculation, but there is reasonable scientific consensus that it began in water. As the young earth cooled, water condensed from its steamy atmosphere to form oceans on the hardening crust. Those oceans, 4000 million years ago, would have been a 'soup' in the form of a weak solution of organic molecules, the building blocks of the living organisms to come. The molecules could have been formed *in situ*, or may have arrived in impacting comets that would also have provided some of the water for the earth's primal ocean. Therein it is usually assumed by biologists that life began, the key questions then being: where in the ocean is it most likely to have occurred and how? What are the natural conditions under which it is possible to visualize that self-replicating, complex molecules, that is primitive living organisms, might have arisen? A common theme in the speculations as to how that happened focusses on the need to postulate a mechanism whereby the primal soup could have been concentrated. This would be a critical requirement to encourage the aggregation of organic molecules by the process of polymerization into more complicated, self-replicating molecular structures that characterize living systems. Charles Darwin (1859) speculated that life probably began in 'some warm little pond with all sorts of ammonia and phosphoric salts' and modern views have developed along similar lines, despite the fact that reservations have been expressed concerning Darwin's original idea. Chemists were quick to point out that the process of polymerization, whereby simple organic molecules are transformed into large complex ones, cannot be achieved in an aqueous medium, prompting the comment in popular science literature that 'if you make monomers wet they don't turn into polymers – except when creating life on earth' (Bryson, 2003). However, maybe the monomers were not permanently wet. Perhaps life began in tidal pools in which the 'primal soup' was subjected to polymerization by intermittent concentration and drying out under the sun. Pools above the level of neap high tides would not only have been subjected to drying out twice each day during spring tides, but would also have been continuously exposed to the atmosphere for several days at a time during neap tides.

The best contemporary laboratory models of pre-cellular systems are liposomes, artificially constructed spheres of fatty material that enclose an aqueous solution. Encapsulation of the self-replicating molecules of DNA within liposomes has been achieved in laboratory cultures subjected to cycles of dehydration and hydration similar to those occurring in pools near high water mark on tidal shores (see Benn, 2001). As noted in Chapter 1, living organisms in coastal zones have clearly been exposed for millions of years to lunar influences, directly, or indirectly through moon-induced ocean tides. Moreover, at the time of the origin of life the moon was probably much closer to the earth than at present. Accordingly, tidal ranges would have been possibly 20% larger than at the present time, revealing particularly large areas of tidal pools that would be subjected to cycles of wetting and drying. With a correspondingly shorter lunar orbit of the earth the neap tide/spring tide interval would also have been shorter than at present. However, large tidal excursions combined with neaps/springs and seasonal effects could well have generated extensive cycles of wetting and drying seemingly necessary for polymerization to have occurred. If so it is not unreasonable to conclude that the moon may have been directly involved in the origin of life itself.

The creation of living matter in a test-tube is not yet a reality, not surprisingly for those for whom the 'notion of an infinitesimally unlikely series of chemical reactions . . . remains the unbidden and silent observer at much of the discussion of how life originated' (Conway Morris, 2003). However, evidence thus far returns us to a version of the Darwinian notion of 'primal soup' with the physical intervention of ocean tides driven by the pull of lunar gravity.

However life began it can reasonably be stated that, once it did so, lunar influences continued to affect the evolution of single-celled and then multicellular, largely soft-bodied, organisms in warm, shallow tidal waters for three billion years or so up to the present day.

Soft-bodied animals would have predominated through the Pre-Cambrian to the Cambrian era, about 500 million years ago, before more complex animals with skeletal tissues evolved and began to colonize the tidal zone, on their way to the exploitation of land. As was seen in Chapter 1, the lunar month and tidal interval would have been shorter in Cambrian times than those of the present day but tidal, and hence lunar, influences on organisms in the coastal zones of the world's seas have clearly been pervasive for aeons of time.

Equally pervasive is the perception by humans, probably going back to the dawn of human thinking, that lunar phases affect the behaviour and physiology of men and women. Even as late as the early nineteenth century in Britain the influence of moonlight was apparent on significant aspects of human behaviour, including members of the Lunar Society of Birmingham who met regularly on the Monday nearest to the full moon over a period from 1776 to 1820 (Uglow, 2002). The members were in no way part of the lunatic fringe, but were highly respected scientists, engineers and industrialists, including Erasmus Darwin, grandfather of Charles, and Josiah Wedgewood. Their objective was to enjoy mutual intellectual stimulus as they sought to advance the cause of their respective disciplines and entrepreneurial endeavours. They chose to meet during the times of full moon, for no other reason than it ensured they had sufficient light at night for a safe journey home after their scientific exchanges. Members of the Society were highly influential during the Industrial Revolution in Britain over the period 1760–1840, and it is ironic to note that they were probably directly responsible for the subsequent development of urban street lighting, which eliminated the reason for the timing of their meetings in the first place. However, though they were directly responsible for the reduction in the significance of moonlight as a source of nocturnal illumination in towns and cities today, the moon is still considered by many to affect human life in modern times. As recently as 1995 a questionnaire sent to 325 people in New Orleans, USA, concluded that 140 individuals (43% of the sample) held the opinion that lunar phenomena alter human behaviour (Zanchin, 2001). Specifically, it emerged that mental health professionals (social workers, clinical psychologists and nursing aides) held this belief more strongly than did other professional groups. This is perhaps not surprising when Charles Darwin himself, commenting on the near coincidence of the lunar cycle and the human menstrual cycle, asked 'if man is descended from fish, why should not the 28-day feminine cycle be a vestige of the past when life depended on the tides, and therefore on the moon?'. At present there is insufficient hard evidence for assuming any causal relationship between the lunar cycle and the human body, yet the possibility of lunar effects on human physiology and pathology continues to obtain popular support, and even to stimulate research by some scientists (Zanchin, 2001). It is a field in which the Lamont Cole's unicorn may safely graze, generating scepticism as it does so in most of the scientific community. The problem is that the scepticism sometimes spills over into a reluctance to accept that there may be any causal links between the moon and aspects of animal behaviour in general.

Fifty years after it was published in the mid-twentieth century, a paper that had long disappeared from the scientific literature was still circulating on the Internet with conclusions said to be factual. The article purported to show that oysters transported from the east coast of the USA to an inland laboratory in Illinois rephased their tidal rhythms to correspond with the times of lunar zenith in their new locality (Brown, 1954). The author claimed the results to show that the oysters were responding directly to daily changes in lunar gravity, a claim still being taken as true by surfers of the Internet early in the twenty-first century. In fact, as Enright (1974) pointed out, if the claim was justified it 'would represent one of the most exciting sensory phenomena ever to be documented', yet no-one has sought to repeat the experiment and bask in the glory of the discovery. The lack of interest and the disappearance of the paper from scientific view no doubt relate to the fact that re-analysis of the original data questioned whether the oysters showed any rhythm at all, let alone one phased to the moon (Enright, 1965b). If there was scepticism about the influence of the moon on living systems before exchanges such as this in the scientific literature, it is not surprising that there was even greater scientific scepticism afterwards.

In order to challenge scepticism concerning the existence of lunar periodicity in living organisms the way forward was to move on from anecdotal accounts, replacing them with sound observations and evidence that could generate tractable questions and hypotheses to be tested by further observations and by sound, repeatable experiments. Encouragingly, such an approach had already been carried out in the early twentieth century, following reports of a fortnightly rhythm of sexual reproduction in the brown alga *Dictyota dichotoma* on British shores. Production of sexual sori during the summer breeding season of this alga was observed to be cyclical, corresponding to the interval between spring tides and shown to be maintained in specimens that had been removed from the influence of tides (Williams 1898, 1905; Müller, 1962). Accordingly, by the mid-twentieth century the questions being asked by a few marine biologists were whether it was possible to show experimentally that aspects of the behaviour of coastal organisms could be related causally to lunar phase, either indirectly through tidal influences, or directly, say, by moon-light. Surprisingly, to some, the answer to both these questions turned out to be in the affirmative.

In recent decades lunar-related behaviour, particularly spawning, has been characterized in a wide range of marine organisms from algae, sponges and corals, through worms, molluscs, crustaceans and echino-derms, to fish. A classical example of the phenomenon in marine animals, described in detail in Chapter 6, relates to the grunion *Leuresthes tenuis*, a small atherinid fish which deposits its eggs in the sand of beaches in southern California. Experimental studies have yet to be carried out on the grunion, but a reasonable working hypothesis is that its behaviour is partially under biological-clock control and is phased by the tides rather than directly by the moon. However, marine fishes with which experi-mental studies of semilunar periodicity have been carried out are two species of the killifish, *Fundulus grandis* and *F. heteroclitus* (Hsiao and Meier 1989, 1992). Fish from a normal habitat of killifish from the Gulf coast of south eastern USA were held under standard laboratory conditions over four to five months during the summer breeding season and mon-itored for evidence of spawning activity. Daily collections of eggs were made to reveal that *F. grandis* and *F. heteroclitus* each exhibited a semilu-nar cycle of spawning activity which free-ran with periods of 13.7 and 14.8 days, respectively. In the unlikely event that the fish were influ-enced by cues external to the laboratory it was concluded that both species exhibit endogenous, temperature-independent, semilunar spawn-ing rhythms of very precise periodicities. Moreover, semilunar variations in the patterns of production of daily hormonal rhythms in preparation for spawning have been demonstrated in *F. grandis* (Emata *et al.*, 1991). In seeking to discern whether the rhythms were phased by the moon directly or indirectly through the neaps/springs cycle of tides Hsiao and Meier (1989) noted that the period of the spawning rhythm of *F. heteroclitus* (14.8 days) was more or less exactly half of the duration of the sidereal month (29.5 days) (see Chapter 1), the interval between successive full moons as viewed from the earth. Peak spawning occurred several days after new and full moons and several days after maximum spring tides. Yet, despite the fact that spawning follows the twice-monthly changes in tidal height determined by the lunar cycle along the US Atlantic coastline where the fish occur, the authors were unable to assess whether moonlight or tidal height was the likely zeitgeber for the reproductive rhythm. However, the spawning rhythm of *F. grandis* collected from the coast of the Gulf of Mexico, with a cycle of 13.7 days, was more closely in synchrony with the synodic month (27.3 days)(see Chapter 1). Atypically in the Gulf of Mexico the unusual neaps/springs cycle of one high tide each day parallels the synodic month, not the sidereal month, so *F. grandis* spawning appears to be sychronized by changes in tidal height and not by changes in moonlight, which follow the sidereal cycle. In this case semilunar rhythmicity appears clearly to be phased by tidal height.

Another example of moon-related rhythmicity in marine organisms which has long been well known anecdotally in maritime human communities concerns the gastronomic quality of shellfish which often varies with the phase of the moon. Indeed, Aristotle stated that in Mediterranean edible sea-urchins (probably *Echinus acutus*) the roe was largest at times of the full moon. Despite this authoritative, albeit historical, statement it was not until about 2300 years later, as Morgan (2001) has pointed out, that Munro Fox (1923, 1928) saw in Aristotle's observation a testable hypothesis. By counting the number of sea urchins of the species *Diadema setosum* with mature gonads over an extended period of time it was possible to demonstrate a distinct reproductive cycle, during which the sea urchins spawned at times of the full moon. The correlation between spawning and lunar phase was convincing, but the causal mechanism between the two, if there is one, was not clear. Nevertheless the hard evidence of lunar periodicity of spawning justifies further investigation. Certainly one can envisage the evolutionary advantage of synchronized spawning in organisms such as sea urchins that release male and female gametes into the sea. Fertilization and the development of free-swimming larvae occur externally before the larvae metamorphose and settle on the sea bed again as miniature adults. Reproductive success and survivorship of the species would be enhanced if natural selection favoured close synchronization of the breeding times of males and females. But the biological advantage of spawning at the times of full moon and not at another lunar phase is not immediately clear in a Mediterranean species, as is also the case concerning lunar periodicity of gamete release in the brown alga *Fucus vesiculosus* in the tideless Baltic Sea (Andersson *et al.*, 1994).

Moreover, if the advantages of spawning at the times of full moon are unclear for a species such as *E. acutus* living in the virtually tideless Mediterranean, they are equally difficult to understand in full moon-spawning sea urchins which live in tidal localities. If it is of advantage for such sea urchins to spawn during the full moon spring tides, then why not spawn every two weeks, not monthly, by also doing so at the new moon spring tides? Is it a tidally related phenomenon or is spawning again directly related to lunar phase? On the face of it, since spawning does not occur during each period of spring tides one might assume that timing is in some way directly related to the phase of the moon as appears to be so in the case of the Mediterranean species. However, as was noted in Chapter 1, it has to be remembered that the height of spring tides varies between new and full moon phases, with largest spring tides alternating between new and full moon spring tides every seven months or so, on the fourteen-month syzygy cycle which affects most coastal locations (Dronkers, 1964). Is it the case, then, that monthly spawning marine organisms in tidal localities do so only on the larger (or smaller) of the alternate spring tides?

These questions were addressed very specifically by Kennedy and Pearse (1975) and by Skov et al. (2005). The former authors studied the monthly spawning behaviour of the sea urchin Centrostephanus coronatus living around Santa Catalina Island, California. They compared the spawning times of urchins collected in two summers, one in which the more extreme spring tides coincided with the new moon and the other when extreme spring tides coincided with the full moon. This species, unlike some others, was found to spawn, not at full moon, but around the time of the third lunar quarter. However, just as the highest spring tides alternate between new and full moon on a seven-month cycle, consequentially the least extensive neap tides alternate between the first and third lunar quarters with the same periodicity. Despite the difference in neap tidal range between the two years of study, Centrostephanus spawned at the times of the third lunar quarter in each year, and not at the time of the smaller (or larger) neap tides. For Centrostephanus, at least, the monthly spawning rhythm appears to be synchronized directly by changes in moonlight (Kennedy and Pearse, 1975). This conclusion is not to say that all sea urchins show monthly rhythms of spawning, nor indeed that all the ones that do so are cued by the light of the moon. It is the case that rhythmic spawning of this kind occurs in only a few species of sea urchins anyway. Moreover, within any one species that does show moon-related spawning, the timings may vary between populations living in geographically separated localities, a phenomenon which also occurs in some corals (Ward, 1992; Permata et al., 2000). So, synchronization to a specific lunar phase seems unlikely to be of particular advantage.

In contrast, several species of crab, including species of Uca living high in the intertidal zone of east African mangrove beaches, were observed to synchronize their lunar larval-release behaviour, sometimes with the full moon and sometimes with the new moon (Skov et al., 2005). Most of the populations of crabs studied released larvae during maximal spring tides when such times occurred during full moon phases, but changing to new moon timing of larval release at those times as a result of switching during the 14-month syzygy tidal cycle. The tendency to release larvae on the highest spring tides increased the higher the crab population lived on the shore. None of the crabs migrated to the water's edge to release larvae, but did so from the shelter of burrows or crevices between tidemarks. Hence high shore populations of the crabs were compelled to release larvae during the highest spring tides (Skov et al., 2005), as has also been noted for other crab species (Morgan and Christy, 1995), gastropod molluscs (Berry, 1986) and fish (Hsiao et al., 1994). Larval release at times of high spring tides for all these organisms appears to facilitate rapid offshore dispersal during strong ebb tides, argued to be advantageous in avoiding confrontation with high densities of planktivorous fish in coastal localities (Johannes, 1978; Morgan and Christy, 1994, 1997).

The reciprocal of tide-related spawning in high-shore species is that organisms on the low shore, in the subtidal or in non-tidal localities seem to be more dependent for their lunar periodicity upon direct control by moonlight. Synchronized spawning, however it is achieved, no doubt maximizes the chances of egg fertilization (Olive et al., 2000); it enhances the chances of larval survival by 'swamping' and improved 'safety in numbers' under predatory pressure (Johannes, 1978; Karban, 1982). There are therefore selective advantages for lunar rhythmicity to have evolved in localities where the moon offers reliable cues for synchronization (Palmer, 1995a; Morgan, 2001). Preference for larval release during new moon might be because, in fish, the lack of moonlight reduces the risk of predation by visual feeders (Johannes, 1978) and spawning by the light of the full moon might facilitate migration of adults to a spawning site (Colin et al., 1987). However, in general, the adaptive advantage of synchronization to a particular lunar phase remains enigmatic.

In contrast to biorhythms of lunar periodicity, which may or may not be externally cued by lunar phase or tidal state, many organisms exhibit rhythms of semilunar periodicity which, intuitively, seem most likely to be phased by tidal state. For example, many low shore-living crabs exhibit such two-weekly rhythms of larval release (Morgan and Christy, 1995), which are mainly synchronized with tidal amplitude (Morgan, 1996). In addition, oysters in various locations are known to 'spawn' on a fortnightly basis, at the times of waxing and waning moon (Korringa, 1957). In these cases, as in crabs, 'spawning' should more accurately be described as 'larval release' since fertilized eggs are brooded amongst the gills of the parental oyster for several days before being released as planktonic larvae. Brooded eggs are fertilized with sperm brought in by the inhalant respiratory current and are released as swimming larvae during neap tides. It is assumed that the occurrence of weak currents during neap tides reduces the risk of excessively wide dispersal of larvae away from the favourable substratum of the parental oyster beds, which are often localized. Similar larval retention mechanisms have also been postulated as a basis for reproductive success in some reef corals. For many corals the extent of larval dispersal remains largely unknown, but the few species for which data exist seem to produce larvae in ways that reduce the chance of transportation away from the parental reef. Some corals have been found to produce buoyant eggs which are distributed locally over the parent reefs, but which are not scattered randomly in ocean currents. Spawning is restricted to the times of neap tides when tidal flushing is least likely to disperse eggs and larvae away from the reef system (Hughes, 1983). Random dispersal of larvae in strong currents during new and full moon tides would be more wasteful than release during weaker neap tides. Excessively wide dispersal would be an evolutionary disadvantage for corals in reefs that fringe oceanic islands. These may be considerable distances from suitable settlement sites elsewhere and separated from them by vast stretches of deep ocean which would be a hostile environment for the larvae of reef corals. On the other hand, some corals, and other coastal organisms, spawn during new or full moon phases, thus releasing larvae at times of spring tides. The advantage of this is considered to facilitate rapid offshore dispersal away from coastal areas where the density of planktivorous fish tends to be high (Johannes, 1978; Morgan and Christy, 1994, 1997).

Interpretation as to how the evolution of some semilunar patterns of spawning may have come about is assisted by studies of bluehead wrasse *Thalassoma bifasciatum*, a common coral reef fish found throughout the Caribbean. Adults of this species spawn on the reefs, after which larvae hatch from the fertilized eggs and are distributed widely in the open sea. Eventually fully developed larvae return to the reefs and burrow in sand for a few days before they metamorphose and emerge as juveniles, which recruit back into the parent population. In a comprehensive study of *Thalassoma* carried out over a 20-month period it was shown that that recruitment of young fish to the parental reefs occurred in a cyclical pattern, most recruits appearing during neap tides, coincident with the first and third lunar quarters (Sponaugle and Pinkard, 2004). Perhaps surprisingly, however, the authors also found that spawning behaviour of the parental fish was not similarly cyclical. During most of the 20-month study adult fish were observed to spawn around midday or early afternoon every day of the month at most of the reef locations visited. So, the semilunar pattern of recruitment of young fish was not the consequence of a semilunar pattern of spawning but was presumably determined by events that influenced larvae during their period of development in the open sea. There are sporadic events such as large swirls and eddies generated by discharging rivers or atmospheric storms, causing what might be called 'ocean weather', but the most consistent pattern of oceanic events relates to alternating strengths of tidal flow between spring tides and neaps. Accordingly it is not unreasonable to consider that bluehead wrasse juveniles recruit back to the parental reefs most successfully during gentle tides associated with neaps, because during spring tides strong currents would tend to disperse them far and wide, most never to find a suitable site for settlement and metamorphosis (Sponaugle and Pinkard, 2004).

Yet another example of lunar periodicity was referred to in Chapter 4, concerning studies in New Zealand which showed that crab larvae returning from the open sea to settle in the intertidal zone are attracted to the sound of waves breaking on coastal reefs, particularly during neap tides, at the times of the first and third lunar quarters (Jeffs et al., 2003). On the coasts in question, the timing of tides is such that the rising neap tides occur immediately after nightfall, the time when free swimming larvae and other planktonic animals would normally be expected to swim upwards towards the sea surface as part of their daily pattern of vertical migration (see Chapter 7). In that particular study the crab larvae swam up from the bottom during flooding neap tides, thus exploiting onshore currents that took them to their preferred coastal sites for settlement.

So, there are oceanic conditions that appear to favour semilunar rhythmicity in the reproductive behaviour in some marine animals, just as there are examples of lunar breeding periodicity known since the time of Aristotle. The question now is whether such rhythms are amenable to experimental study. Evidence is required in the quest to understand whether they are controlled by internal biological clocks in the organisms concerned and, if so, whether their lunar timing is achieved indirectly through tidal cues or directly through the influence of the moon itself.

*   *   *   *

Lunar and fortnightly rhythms of behaviour that have been studied experimentally occur in a number of marine crustaceans (Naylor, 2001). Some of these rhythms, for example the larval release rhythms of crabs (DeCoursey, 1983; Forward, 1988), have been shown to be endogenous, persisting in constant conditions, and should therefore be designated

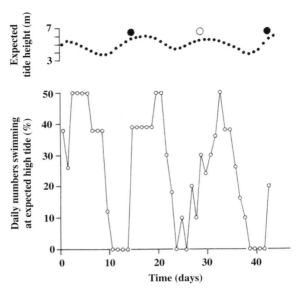

Figure 5.1 Daily percentage numbers swimming at times of expected high tide of 40 *Eurydice pulchra* freshly collected each day for 43 days and kept in constant conditions for 24 h. Open circles, full moon; closed circles, new moon (redrawn from Alheit and Naylor, 1976).

circalunar and circasemilunar, as defined in Chapter 1. The sand-beach isopod crustacean *Eurydice pulchra* again provides a good example to illustrate this point. As seen in earlier chapters this isopod emerges from intertidal sand to swim and feed in large numbers as the tide rises, re-burrowing as the tide falls. In constant conditions in the laboratory the animals show a circatidal rhythm of swimming at the times when they expect high tide. These crustaceans survive well in laboratory aquaria and it is under such conditions that the isopods also show fortnightly increases in the magnitude of their bursts of swimming. Without being able to see the moon isopods freshly collected each day emerge from sand at the bottom of the tanks and swim in large numbers just after the times of full and new moon, but in fewer numbers, if at all, at the times of lunar quarters (Figure 5.1) (Alheit and Naylor, 1976). Moreover, when isopods were freshly collected during neap and spring tides and maintained in constant conditions in the laboratory for two to three days, both showed a circatidal rhythm of emergence and swimming at the times of expected high tide, but those collected during spring tides swam in far greater numbers than those collected during neaps (Figure 5.2) (Reid and Naylor, 1986). In their normal habitat on the home beach such spontaneous behaviour helps to ensure that maximum

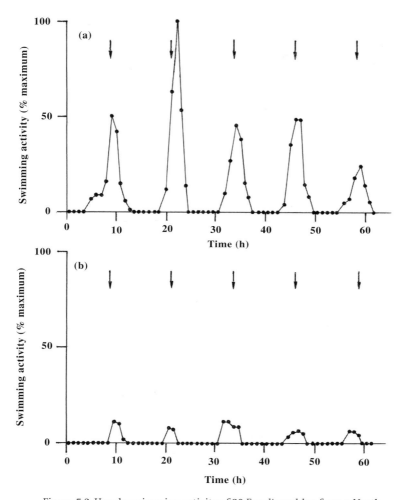

Figure 5.2 Hourly swimming activity of 20 *Eurydice pulchra* from a North Wales beach, recorded in constant conditions over five expected tidal cycles after capture: (a) at spring tides; (b) at neap tides. Values are percentage maximum related to the highest hourly value recorded during spring tides. Arrows, expected times of high tide (redrawn from Reid and Naylor, 1986).

swimming of the whole population of *Eurydice* occurs just after times of the highest spring tides when tidal oscillations are declining from their maximal range. It appears that such precisely timed swimming en masse enables *Eurydice pulchra* to move slightly downshore during the ebbs of reducing spring tides and so avoid being stranded in dry sand above high water mark during the subsequent neap tides (Alheit and Naylor, 1976).

If this interpretation of *Eurydice* behaviour throughout the neaps/springs cycle is correct it implies that the isopods have not only a circatidal biological clockwork, as described in an earlier chapter, but circasemilunar (approximately 14 day) biological clockwork too. At least, evidence so far suggested a hypothesis to that effect which was amenable to testing. For example, is it possible to manipulate experimentally the behavioural rhythmicity of *Eurydice* with a view to finding out how such a circasemilunar rhythm might be regulated by environmental cues?

Investigation of this problem began with the questions as to whether the timing might in some way be directly related to moonlight or, alternatively, indirectly related to lunar phase through the influence of tides (Reid and Naylor, 1985, 1986). Since maximal swimming occurred during spring tides associated with both new and full moon, as indicated earlier, it seemed reasonable to assume, intuitively, that tidal factors, not lunar factors directly, were critical in semilunar timing. Moreover, around the British Isles, where the study was carried out, weather patterns and irregular cloud cover make direct lunar cues unreliable. On this basis, moonlight seemed to be an unlikely synchronizer and the most economical initial hypothesis to test was that which postulated control by tidal variations associated with the neaps/springs cycle.

Beginning with the finding established by Hastings (1981a) that the phase-setter for the circasemilunar rhythm of *Eurydice pulchra* was the timing of high tides in relation to the time of the solar day, studies were first made of isopods collected from a beach in North Wales where high spring tides occur around noon and midnight (Reid and Naylor, 1985). Specimens were taken into the laboratory and entrained for four days in an artificial tidal regime in which they were exposed to water agitation simulating intense wave action for two hours around noon and again around midnight. After this treatment the isopods were maintained in constant conditions in the laboratory and their total daily swimming activity was recorded every day for 50 days or so. Every time an isopod swam through a beam of infra-red light projected across their tank a digital counter summed their daily total on a remote computer screen. During that time, in the absence of any external time cues, they exhibited a rhythm of swimming intensity, with peak swimming occurring at intervals of about 14 days. The peak swimming times occurred on the 14-day anniversaries of the timing of their last exposure to an episode of simulated tidal action (Figure 5.3) (Reid and Naylor, 1985). This finding provided evidence that the artificial tides had indeed initiated a circasemilunar rhythm of swimming of exactly the pattern exhibited by *Eurydice* when brought into the laboratory fresh from the beach

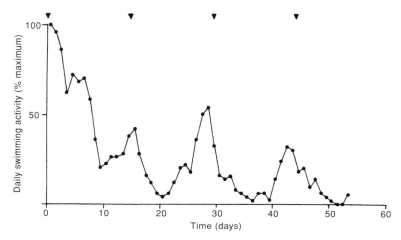

Figure 5.3 Induced rhythm of semilunar periodicity in 20 *Eurydice pulchra* entrained by artificial tidal agitation for 2 h every 12 h for 4 days, after which swimming activity was recorded continuously for 54 days in constant conditions. Total daily swimming activity is plotted as percentage maximum, for each day in constant conditions; arrows, semilunar (approximately 14 day) intervals from the end of entrainment (redrawn from Reid and Naylor, 1985).

(Figure 5.1). Further evidence supporting this conclusion was then obtained by subjecting isopods to four days of exposure to artificial tides at different times of day. When recorded subsequently in constant conditions, specimens that had been exposed to artificial tidal agitation around noon and midnight showed far greater circatidal swimming activity than a similar group of animals that had been exposed to artificial tidal agitation around times of dusk and dawn (Figure 5.4), closely resembling the pattern exhibited by freshly recorded isopods at springs and neaps, respectively (Figure 5.2) (Reid and Naylor, 1986). So the initial hypothesis appeared to be validated, with evidence not only of endogenous twice-monthly swimming behaviour in the isopods, but that the animals were specifically adapted to semilunar cues related to the timing of tides on their native beach. Needless to say, the findings do not necessarily rule out the possibility of entrainment by monthly variations in the intensity of moonlight but that possibility seemed even less likely when a comparative study was made of a population of the isopods from another beach with a different tidal regime.

This additional study was carried out using *Eurydice pulchra* from a beach in South Wales where high (and low) tides occur around six

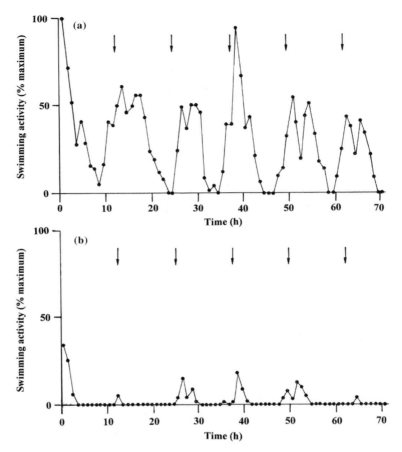

Figure 5.4 Hourly swimming activity of 20 *Eurydice pulchra* from North Wales recorded in constant conditions after exposure to an artificial tidal entrainment regime of 2 h agitation every 12 h for 4 days: (a) at midday and midnight (spring high tide timing); and (b) at dawn and dusk (neap high tide timing). Values expressed as percentages of the highest value in (a); arrows, 12.4 h intervals from end of entrainment (redrawn from Reid and Naylor, 1986).

hours earlier than in North Wales. On the South Wales beach high spring tides occur around dawn and dusk, and low spring tides occur around noon and midnight. Isopods from the South Wales beach could also be induced to show a persistent circasemilunar swimming rhythm in constant conditions in the laboratory, but they only did so after exposure to artificial tidal regimes with maximal water agitation at the times of dawn and dusk, and not at the times of midnight and midday (Reid and

Naylor, 1986). The two populations of *Eurydice pulchra* therefore behave in an exactly complementary way when exposed to the tides (Figure 5.5), indicating localized adaptation by members of the same species over the relatively short distance of coastline. As discussed in Chapter 1, the delay in timing of tides between the South and North Wales beaches arises because the tidal bulge takes about six hours to travel northwards over the distance separating the two study localities, entirely within the range of distribution of a single species such as *Eurydice pulchra*. Tidal variations may therefore present a range of environmental characteristics within the geographical spread of individual species of animals in the inter-tidal zone. It is then not surprising that though evolutionary adaptation has selected for biological clock mechanisms that permit anticipation of tidal oscillations, it has also favoured mechanisms that permit flexible synchronization of biological clocks to local tidal regimes, characteristics that must be taken into account when considering the nature of internal clock mechanisms.

*   *   *   *

More studies are required to demonstrate unequivocally that lunar and semilunar rhythms of behaviour in marine organisms are generally of the *circa*-variety, which persist under biological clock control in the labo-ratory. There is mounting observational evidence of lunar and semilunar patterns of behaviour in a wide range of organisms in the sea, but only in relatively few species have critical experimental studies been carried out to determine whether the rhythms are controlled by true circalunar or circasemilunar biological clocks. In some circumstances semilunar rhythms could be purely exogenous in origin or they could be gener-ated by the 'beat effect' of interacting circadian and circatidal oscillators whose maximal oscillations would coincide every 14–15 days or so. Fort-nightly rhythms of locomotor activity such as those in *Eurydice pulchra* have been reported in a related isopod *Excirolana chiltoni* on Californian beaches (Enright, 1972; Klapow, 1976) and in the sand hopper *Talitrus saltator* on British beaches (Williams, 1979). Also, as will be discussed in Chapter 8, various authors have reported for estuarine animals fort-nightly patterns of swimming behaviour timed to rising spring tides, as mechanisms ensuring retention in the estuary and avoidance of being swept out to sea.

Semilunar patterns of moulting have also been demonstrated and shown to be of adaptive advantage in two species of crustaceans which live high in the intertidal zone, including the semi-terrestrial amphipod *Talitrus saltator* which moults with greatest frequency 5–7 days before the new and full moons. In *Talitrus*, moulting times coincide with neap tides,

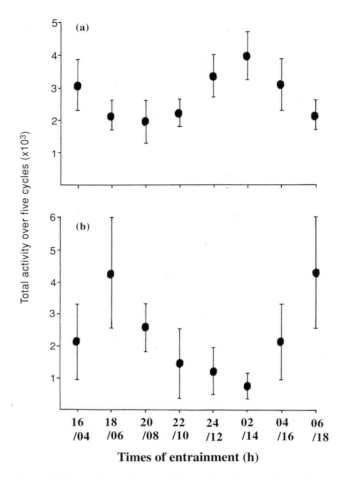

Figure 5.5 Mean (± standard deviation) of total swimming activity over five tidal cycles in constant conditions for groups of *Eurydice pulchra* after exposure to various regimes of 2 h artificial tidal agitation every 12 h for 4 days, plotted against timings of tidal pulse application, which were initiated at 04.00 h and 16.00 h, 06.00 h and 18.00 h, etc. The isopods were collected from populations on beaches in: (a) North Wales (high spring tides at noon and midnight); and (b) South Wales (high spring tides at dawn and dusk). Last two points are repeated in each case (after Reid and Naylor, 1986; by permission of Elsevier).

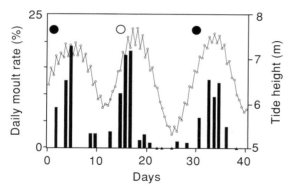

Figure 5.6 Percentage of juvenile shore crabs *Carcinus maenas* moulting within 24 h of collection throughout the lunar cycle. Maximum daily tidal height on the collection shore is also plotted. Triangles, no moulting; open circle, full moon; closed circle, new moon (after Zeng *et al.*, 1999; by permission of Inter-Research).

when newly moulted hoppers would be least mobile and high tides would not normally reach their burrowing zone high on the beach (Williams, 1979). Fragile, newly moulted hoppers during neap tides would be less exposed to the potentially damaging effects of wave action than if they moulted during spring tides. Conversely, newly settled young crabs of the species *Carcinus maenas* were found to moult most abundantly around the times of spring tides, just after the times of new and full moon (Figure 5.6) (Zeng *et al.*, 1999). The young crabs live among gravel and stones near high water mark on rocky coasts and they moult when they would normally be covered by seawater at high spring tides. It would be maladaptive for soft-shelled, early juveniles of this marine crab to moult during neap tides, living as they do so high on the shore. They might not be covered by seawater for an extended period of time and in the soft-shelled state after moulting they would risk fatal desiccation.

In fact, crabs and lobsters provide many examples of lunar and semilunar rhythmic behaviour, particularly concerning their reproduction and early life history. On beaches in the tropics and subtropics, fiddler crabs of the genus *Uca* often abound at the mouths of their burrows, frantically signalling to each other with their fiddle-like claws. As indicated earlier, many occur between tidemarks on sheltered beaches, but some extend well into estuaries, marshes and mangrove swamps. Amphibious and often tolerant of freshwater as those species are as adults, their larvae are fully marine and must undertake their early plankton development in the ocean. Eggs carried by ovigerous females must be released directly into the sea and several species have been observed to move in large numbers to the water's edge at high spring tides, that is around the times of full and new moons (Christy, 1978; Wheeler, 1978; Bergin, 1981; DeCoursey, 1983). So far, few species have been studied in constant conditions in the laboratory, to determine whether they continue to release larvae at fortnightly intervals, but larval hatching rhythms in such crabs may be truly circasemilunar, as they appear to be in land crabs which have similar requirements to return to the sea to release their larvae (Saigusa, 1980, 1982, 1986). It has not been ascertained whether synchronized larval release at high spring tides enhances the survivorship of adult female fiddler and land crabs, but it is possible that this is so. Energy expenditure and risk of predation in moving to the water's edge from above high tide level would be least at high spring tides when the distance of migration would be shortest. Also there may be additional survivorship advantage to the crabs since newly released larvae would be more effectively dispersed in the stronger offshore tidal currents at spring tides than at neaps. Unlike larvae of some animals that have a short larval life and for which release during weak neap tides is an advantage, crab larvae normally have an extended period of existence in the sea. This permits prolonged feeding in coastal waters where they grow through several moult stages before returning shorewards to settle out as young crabs.

Needless to say, since there is a biological requirement for the larvae of coastal and shore-living adults to undergo early development in the open ocean there must also have been selective advantage in the evolution of reciprocal mechanisms ensuring replenishment of the parental stocks. This raises interesting questions and hypotheses, not only for curiosity-led science, but also for fisheries scientists concerned, for example, with the maintenance of inshore commercial stocks of crabs and lobsters. It has already been noted in this chapter that crab larvae recruiting to New Zealand coasts do so by fortnightly swimming up into the water column, exploiting onshore tidal currents after nightfall during neap tides (Jeffs et al., 2003). Also, similar studies have been carried out on the recruitment back from the open ocean of late larvae and early juveniles of spiny lobsters (see Chapter 9), which form highly valuable coastal fisheries in many parts of the world. Here again it turns out that the moon, directly or indirectly, may have an influence on recruitment in some localities, though not in others. For example, recruitment of larvae into commercial stocks of spiny lobsters in Australia, but not in California, has been shown to vary differentially according to lunar phase. In western Australia last stage larvae of *Panulirus cygnus* settled out from the plankton more abundantly at time of the new moon than during the full moon periods (Phillips, 1977), whereas settlement of *Panulirus interruptus* on Californian coasts showed no correlation with lunar phase (Serfling and Ford, 1975). Whether linkage with the moon is indirect through tides or direct through moonlight intensity continues to be a matter for speculation, but at least observed correlations pose interesting and tractable questions for further study. This leads one to ask whether there is any experimental evidence to support the possibility that moonlight may directly control the behaviour of some marine organisms.

*   *   *   *

Following earlier, even historical, anecdotal accounts of the coincidence between the spawning of marine animals and lunar phase, probably the first comprehensive study indicating direct lunar phasing of a breeding rhythm in a marine animal was carried out by Hauenschild (1960). He worked with the marine polychaete worm *Platynereis dumerilii* at the Stazione Zoologica in Naples, where moonlight might perhaps be regarded as a more reliable environmental cue than on the northern coasts of Hauenschild's native Germany. He first observed that the worms spawned most abundantly at the times of new moon and then, after culturing worms in normal day/night conditions in the laboratory, found that they too spawned at the same time as the natural population in the sea (Figure 5.7). At sexual maturity the worms transform into breeding mode, the stage at which they swarm at the sea surface to spawn and then die.

In worms cultivated in continuous artificial light in the laboratory, Hauenschild found that spawning occurred uniformly throughout the month (Figure 5.7). However, lunar periodicity could be reinstated in an otherwise randomly spawning collection of worms in the laboratory by subjecting them to dim light, equivalent to moonlight, for 12 h during a period of a few consecutive nights (Figure 5.8). Evidently some aspect of this artificial lunar event served as a direct synchronizer of the lunar spawning rhythm. Moreover, the worms did not simply spawn once, as if they were responding solely to their last experience of artificial moonlight, which would be the case if their biological chronometer functioned as a countdown timer. The monthly spawning rhythm persisted in the laboratory culture for three months without further experience of moonlight, real or artificial. Since the rhythm persisted in the laboratory in this way it clearly deserves description as a true circalunar rhythm of approximately 29-day periodicity that is synchronized directly by moonlight. Indeed, Hauenschild (1960) also showed by varying the number of nights during which the worms were exposed to moonlight, that there was a

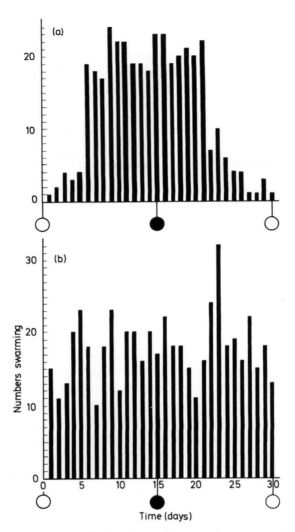

Figure 5.7 Distribution of swarming dates in an average lunar month of *Platynereis dumerilii* maintained for one year in the laboratory: (a) 366 swarming individuals in cultures maintained under natural illumination; (b) 524 swarming individuals in cultures maintained under continuous artificial light. Open circles, full moon; closed circles, new moon (after Hauenschild, 1960, in Naylor, 1976; by permission of Cold Spring Harbor Laboratory).

Number

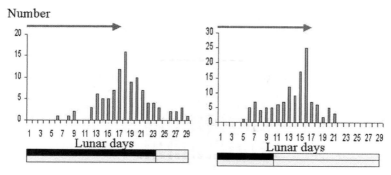

Figure 5.8 Number of swarming dates in an average lunar month of *Platynereis dumerilii* maintained in the laboratory under 12 : 12 h light : dark conditions, with exposure to artificial moonlight during dark intervals for 5 nights and 18 nights, respectively. Arrows indicate an 18 day interval from 'moon off' to peak spawning in each case (after Bentley *et al.*, 2001; redrawn from data in Hauenschild, 1960; by permission of Springer Science and Business Media).

fixed relationship between the night when artificial moonlight ceased ('moon-off') and the timing of peak spawning. This occurred whether the worms were exposed to 5 or 18 nights of artificial moonlight (Figure 5.8). Clearly, moonlight, or lack of it, is not the trigger for spawning, but light reduction after full moon appears to be the synchronizing factor, or zeit-geber, for the endogenous rhythm of circalunar periodicity (Hauenschild, 1960). Subsequently, after Hauenschild demonstrated lunar entrainment of circasemilunar rhythmicity in a marine worm, Muller (1962) and Vielhaben (1963) were able to demonstrate similar entrainment by artificial moonlight of a persistent circasemilunar rhythm of gamete release in the brown alga *Dictyota dichotoma*. Later, too, Saigusa (1980) demonstrated a circasemilunar rhythm of larval release by the semi-terrestrial crab *Sesarma*, in crabs after a short exposure to artificial moonlight when otherwise maintained in constant conditions in the laboratory.

Even more convincingly, perhaps, the role of moonlight in synchronizing a circasemilunar breeding rhythm has been demonstrated in the intertidal midge *Clunio marinus*, one of the rare insects in the sea and its fringing intertidal zone (Neumann, 1965, 1976b, 1987). Having evolved air-breathing systems as they colonized dry land early in evolution, few insects, as adults, have reverted to solving the problem of breathing when immersed in sea water. Some intertidal midges have solved the problem in an unusual way; they spend most of their lives as eggs, larvae and pupae during which time they survive when immersed in seawater by absorbing oxygen across their body surface. The adults, when they emerge from the pupae, must breathe air, and they are able to survive for only a few hours over one period of low tide before they are drowned by the rising tide. During this short interval of time they must achieve mating and egg-laying before they die. When walking on a European rocky shore at low tide one sometimes sees evidence of this in the form of swarms of midges hovering near low water mark when the mating midge rituals are in progress. The flying midges are the males seeking to mate with the non-flying females, which must then lay their eggs before the rising tide floods the crevices where the developing eggs, larvae and pupae survive repeated tidal immersion until swarms of flying males appear again to repeat the life cycle. Since the zone of occurrence of the developing midges is low down the shore, near the level of low water spring tides, it is clear that the mating and egg-laying swarms of the midges only occur during low spring tides, that is around the times of new and full moon. Numerous questions and hypotheses arose from initial observations of the lifestyle of Clunio, and experiments to test the hypotheses were designed by Neumann once he had found it possible to culture midges through their entire life cycle under laboratory conditions.

He began by studying populations from southern Europe. Stock cultures from the coasts of Spain were first maintained in the laboratory in an artificial day/night cycle of 12 h light and 12 h darkness. Under these conditions male midges did emerge and fly in search of flightless females, but they did so quite randomly, showing no pattern of synchronized emergence. However, the semilunar periodicity of emergence that the midges exhibited on the shore could be induced in the laboratory by a simple procedure. A natural semilunar cycle of emergence, flying and mating behaviour occurred after cultures were exposed to dim light (0.3 lux), equivalent to moonlight, during four successive dark phases of the laboratory light/dark cycle, and repeated again after 29 and 58 days, that is for two complete 'lunar' cycles (Figure 5.9) (Neumann, 1965). It is sometimes argued that by exposing animals to cycles of environmental change the experimenter is conveying information not only about the timing of an event, but also about the periodicity of the phenomenon. Such conditioning, it is suggested, shows that a biorhythm is an expression of learning behaviour and has nothing to do with the possession of a biological clock. However, here is an example of midges that have been exposed to artificial lunar cycles which then express breeding rhythms of semilunar periodicity, suggesting indeed that entrainment by the experimenter is conveying information only about phase, not periodicity, to the midges' biological clocks. The semilunar rhythm expressed by the cultures was clearly endogenous, since it was free-running in the absence of

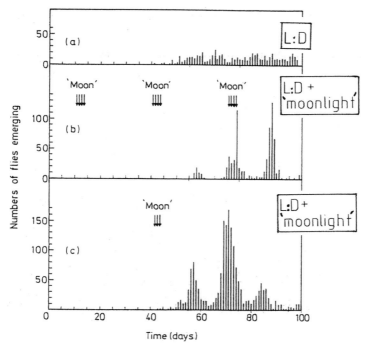

Figure 5.9 (a) Cultures of the intertidal midge *Clunio marinus* in the laboratory hatch randomly when kept in a 24-hour light/dark cycle. In contrast, (b) a semilunar rhythm of hatching is induced by exposure to three 4-night sequences of artificial moonlight at monthly intervals, or (c) even to one 3-night sequence of artificial moonlight (after Neumann, 1965; by permission of Zeitschrift für Naturforschung).

'moonlight' under laboratory conditions and could therefore be regarded as truly circasemilunar. This was confirmed in other experiments in which three nights, or even a single night of exposure to artificial moonlight also reinstated a persistent 14-day rhythm of emergence by adults (Figure 5.9) (Neumann, 1965). In the latter case no information about periodicity was conveyed to the insects, confirming that the cultures reared in the absence of moonlight required information only concerning lunar phase and not about lunar or semilunar periodicity.

Finally, it remained to consider whether lunar-related biorhythms were apparent in the behaviour of intertidal midges in northern latitudes, where variable weather ensures that moonlight is a less reliable environmental cue than on the beaches of southern Europe where Neumann carried out his original experiments. If *Clunio* populations in northern Europe also emerge to fly and mate with semilunar periodicity it is unlikely that they are cued in their behaviour by the unpredictable light of the moon and some other explanation for their behaviour might have to be found. This point was answered by Neumann (1976a), who indeed found that populations of midges in a northern European locality showed a semilunar breeding cycle. Moreover he showed that they did not cue their breeding cycle in response to the moon directly, but indirectly by tidal factors. Laboratory cultures of midges subjected to artificial tidal agitation at particular times of day were found to express a circasemilunar rhythm of breeding behaviour. Moreover he found that there were critical timings of tidal agitation which varied according to the times of spring tides at the locality from which the midges were obtained. He demonstrated quite clearly that geographically separated populations of *Clunio marinus* show genetic adaptations of their endogenous circasemilunar clocks to local tidal patterns. Populations from Helgoland and Normandy both showed maximum emergence during low water of spring tides, but there is a two hour difference in the timing of low water between these two localities. These timings suggested that entrainment of the emergence rhythm should require a different phase-relationship between the time of wave action and the time of day at each locality. This proved to be so, since, when entrained in the laboratory to the same combined cycle of light/dark and simulated wave action, the circasemilunar rhythms of the two populations showed phase differences in the timing of their emergence rhythms. The differences corresponded well with the time difference between afternoon spring low tides in Normandy and on the island of Helgoland, respectively (Neumann, 1976b, 1987).

*   *   *   *

In conclusion, this chapter shows that evidence is building up quite convincingly for the occurrence of moon-related biological clocks in a wide variety of marine organisms. Like tidal and daily clocks they persist away from environmental influences, justifying the *circa* prefix whether expressed as lunar or semilunar behavioural rhythms. The physiological basis of such long-term rhythms has yet to be addressed, and experimental demonstration of their environmental synchronization has only just begun. Lunar phase has been shown to influence directly some animal behaviour patterns and indirectly to affect others, mediated by some aspects of the neaps/springs cycle. But many aspects of the possible effects of the neaps/springs cycle of tides remain unstudied. For example, intertidal beach sediments show semilunar cycles of temperature change, superimposed upon their tidal and daily temperature changes, dependent upon the timing of extreme low spring tides during the solar day (De Wilde and Berghuis, 1979). As discussed in Chapter 1, in any one geographical locality the timing of low spring tides tends to occur repeatedly at about the same time of day. Hence, on some gently shelving beaches, where low spring tides occur around midday (and midnight), changes of beach sand temperature as extensive as 5°C may occur between one low tide and the next, at semilunar intervals. Solar heating of such beaches at low spring tides in summer generates far greater temperature changes over the lunar cycle than would occur in localities where low spring tides occur during dusk and dawn. Possible causal relationships between physical variables such as these and moon-related animal behaviour patterns are worthy of further investigation, as indeed are other examples of lunar periodicity in organisms which are often set in the context of yearly biological rhythmicity. Annual rhythms are the subject of the next chapter which considers the time-course of some fairly precisely timed, yet still often moon-related biological phenomena in the sea which occur at approximately yearly intervals.

6

Annual biorhythms

It thus looks certain that some organisms possess circannual clocks that oscillate, in the absence of environmental time-cues, at a frequency of roughly once per year

John Brady, 1979

It is nearly dawn on 30 October 1980 on the Samoan Islands of Savaii and Upoli (Apia) in the south west Pacific Ocean. The moon is waning and it is the day of its third quarter. According to the Samoan calendar it is a 'Palolo Day'. Palolo is a marine worm, expected to make itself available for capture in huge quantities today, and considered to be a great delicacy by native Samoans. The sense of anticipation among the local fishermen is high, as they wonder if the worms will emerge to spawn, swarming in the sea around the inshore reefs, as the calendar predicts, or whether their preparations will be to no avail. In the event the forecast is correct and the worms suddenly appear in vast numbers, recorded subsequently by scientists of the Fisheries Division of Samoa as a 'heavy spawning strength'. Canoe fishermen were able to scoop up large quantities of the spawning mass of worm tails, providing an annual feast for the local people, as they had done at the equivalent phase of the moon at that time of year for many years previously and have continued to do so since.

The third quarter of the moon does not, of course, always occur on the same days of the year; timings in relation to particular days in a year vary according to a 19-year so-called metonic cycle. So, since 30 October 1980 was the day of a third lunar quarter, the occurrence of a third lunar quarter on 30 October did not occur again until 1999. However, during many years of recordings of the event, spawning has always occurred on or about those third lunar quarter days that occur during a six-week period in October and November. The calendar that specifies this event has now been scientifically established (Caspers, 1984; Naylor, 1985), but it was also well understood for many years previously in native folklore. Annual spawning was precisely anticipated by fishermen in historical times, their canoes and scoop nets being made ready each year in anticipation of the nutritious feast. Only occasionally have they been unrewarded when spawning failed.

The Samoan palolo worm (*Eunice viridis*) spends most of its life in tunnels in coral rocks of reefs that fringe tropical islands. They occur at water depths of 3–5 m and are normally rarely seen. Swarming on Palolo Days is the process by which the worms reproduce. In the weeks before spawning, both male and female worms add numerous additional segments to their bodies, each segment filled with eggs or sperm. Prior to breeding the worms are normally a few centimetres in length but their reproductive 'tails', or epitokes, may extend a further 30 cm, increasing their total body length many times. Spawning is effected by schizogamy, the shedding of epitokes, male and female simultaneously in vast numbers into the water above the parental reefs. The headless tails, or sexual satellites (Bentley *et al.*, 2001), swim up to the surface in massive swarms, eventually releasing eggs and sperm that join in fertilization to produce larvae that grow to replenish the parental stock. It is whilst the epitokes swarm at the sea surface that they are captured by fishermen using scoop nets over the sides of their canoes. Fortunately for the local population, the use of these traditional artisanal fishing methods over the years has permitted sustainable yields of palolo to be taken without over-fishing, a problem which has afflicted other living marine resources around the world that are exploited by industrial-scale fishing methods.

Early biological textbook accounts of the timing of palolo spawning were largely anecdotal, generating scepticism concerning its association with phases of the moon at particular times of year. However, the phenomenon is so unusual that it turns out that the dates of spawning have been quite well documented for over 150 years. Early in that time frame the diaries of European missionaries and medical practitioners on the Samoan islands provide accurate dates of palolo emergence, whilst later visiting biologists and resident fisheries scientists have recorded even more precise timings and strengths of spawning. The first documented record is for 1843 when a third quarter of the moon occurred on 16 October and palolo swarmed on 15/16 October. In 1998, lunar third quarter days occurred on 12 October and 11 November and 'medium to strong' swarming around Upolu and Savaii was recorded by fishery biologists on 11/12 October and 10/11 November, respectively (Caspers, 1984).

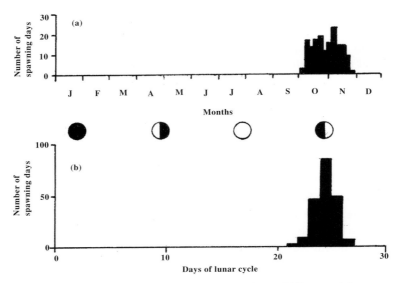

Figure 6.1 Spawning dates of Palolo worms *Eunice viridis* around the Samoan Islands, based on 78 years of discontinuous recorded spawnings from 1843 to 1999, plotted: (a) in relation to days of the calendar year; and (b) in relation to days of the lunar cycle (after Naylor, 2005; data from the Fisheries Division of Samoa; by permission of *Scientia Marina*).

Using 78 years of records kindly provided by the Fisheries Division of Samoa, documented discontinuously from 1843 to 1999, it is possible to pool all recorded spawning dates to illustrate the general pattern of this remarkable biological event (Naylor, 2001, 2005). By plotting all spawning dates grouped in five-day intervals throughout a standard calendar year (Fig. 6.1a) it is apparent that during a century and a half of observations, all Palolo Days occurred during the months of October and November, most occurring within a six-week period during those two months. Then if the same data are plotted in relation to the phases of the moon (Fig. 6.1b) it is apparent that 43% of all spawning days occurred on the day of the third lunar quarter. A further 51% of all spawning events occurred more or less equally on the day before and the day after the third quarter. Only 6% were relatively inaccurately timed, to the extent that they occurred two or three days before or after the third lunar quarter. The precision in relation to lunar phase is therefore quite remarkable.

But the pooling of data in this way obscures the pattern of swarming year on year, bearing in mind the mismatch between lunar and calendar months. This pattern is best visualized by plotting swarming dates in relation to the metonic cycle of 19-year periodicity (Fig. 6.2). Because

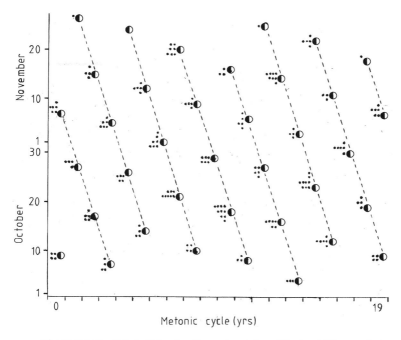

Figure 6.2 The Palolo Calendar: Spawning dates of *Eunice viridis* recorded over a total of 78 years plotted against times of third lunar quarters as they occur over a standard 19-year metonic cycle (after Naylor, 2001; data from the Fisheries Division of Samoa; by permission of Springer Scientific and Business Media).

lunar months are shorter than calendar months, a particular phase of the moon occurs about 10 or 11 days earlier in any one year than in the previous year. As mentioned earlier the progression is such that, if any one phase of the moon occurs on a particular date, the same lunar phase will not occur on the same calendar date until 19 years later. By visualizing all spawning days on a synthetic plot based on the metonic cycle in this way a predictive calendar can be formulated for the months of October and November in the palolo year. There is a 'window' of 6–8 weeks each year when spawning could potentially occur and, if there is one lunar third quarter, during the opening of that window, around the end of October, spawning occurs at that time. If there are two lunar third quarters during the spawning window, with one occurring in early-to-mid October, then the next third quarter in early November is also a potential spawning date. In those years the worms swarm on one or both of the third lunar quarters. Only occasionally does mass spawning fail, for reasons that are still not understood.

The adaptive significance of the remarkable spawning rhythm of the palolo remains largely speculative. Since most spawning occurs around the times of lunar third quarters the larvae are released into the sea at times coincident with neap tides and there may be adaptive advantages in the release of free-swimming larvae during such tides. Neap-tidal currents would be weaker than those swirling around the reefs at spring tides and the weak-swimming, short-lived larvae would be less likely to be swept away from suitable settlement sites on the parental reefs if they were released during neap tides than during springs. Many reef corals also release free-swimming larvae during neap tides and are thought to do so to avoid over-dispersion from the parental coral reefs (Hughes, 1983). As briefly discussed in Chapter 5, wide dispersion such as might occur during more powerful spring tides would increase the risks of transporting larvae to offshore deep waters where favourable settlement sites might not be available. Additional explanation is required, however, to understand why the Pacific palolo, unlike the Atlantic palolo, spawns only on the third quarter neaps and not also during neap tides of the first lunar quarter (Clark and Hess, 1940). Around the Samoan Islands, as elsewhere around the world, there are inequalities of tidal amplitude between successive episodes of neap tides, but these do not help to explain the observed pattern of Pacific palolo spawning. As described in Chapters 1 and 5, in some years the third lunar quarter neaps are smaller in amplitude than those of the first quarter and in other years the opposite is true, but it is clear that these differences have no obvious effect on palolo spawning dates.

Perhaps, as discussed in the previous chapter concerning the nuptial dances of the related marine worm *Platynereis dumerilii* at Naples, some aspect of the lunar cycle itself has a role in palolo spawning. This question was first addressed by Clark and Hess (1940) who studied the swarming behaviour of the Atlantic palolo *Eunice schemacephala*. This West Indian species lives in rock and coral crevices below low tide mark and it swarms at the times of the first and third lunar quarters in July. At those times of year, before dawn, the worms back out of their burrows and the elongated sexual epitokes break free, swimming in a spiral motion to the sea surface. By dawn the sea surface is covered with swarms of detached worm tails which then rupture to release eggs and sperm that fuse to effect fertilization and produce millions of tiny swimming larvae. Within three days the larvae disappear from the water column, sinking to the seabed to restart the worm's life cycle amongst rocks and coral. It was once thought that the spawning of Atlantic palolo was solely in response to moonlight. It was shown by experiments that levels of illumination equivalent to full moonlight, or greater, induced an avoidance response in the adult worms, ensuring that they remained in their coral rock tunnels, whereas the light of a new moon was insufficient to induce any behavioural response whatsoever. Light intensities equivalent to those of the lunar quarters were insufficient to cause the adult worms to retreat into their tunnels, but were sufficiently intense to induce release of the epitokes, which were then attracted to moonlight and therefore swarmed at the sea surface for spawning (Clark and Hess, 1940). However, it was subsequently pointed out by Clark (1965) that this explanation has drawbacks since there are regional differences in spawning times and, in any case, spawning times remained unaffected even when the moon was obscured on cloudy nights. In the Atlantic palolo, and certainly in the Pacific palolo, which spawns on only one of the lunar quarters, it seems likely that spawning is not timed simply by exogenous factors but that internal, endogenous, biological clock mechanisms may also be involved.

As a testable hypothesis, one can suggest that Samoan palolo worms have an annual reproductive cycle during which they have the potential to release larvae during October and November, the southern 'spring' in so far as tropical islands only 17 degrees south of the equator may be considered to have such a season. Whether the annual breeding cycle would persist in constant conditions, defining it as a truly circa-annual rhythm, has yet to be determined. But it is likely to be so, as has been shown for *Arenicola marina* the common lugworm of British shores which, when brought into the laboratory ahead of the spawning season and maintained in isolation under constant conditions of light and temperature, away from the influence of tides, spawns synchronously with conspecific lugworms on their home beach (Howie, 1984). The occurrence of a circa-annual physiological rhythm of maturation and spawning, combined with a hierarchy of responses to falling temperature in autumn and to the subsequent pattern of low spring tides, appear to determine the precise timing of *Arenicola* annual reproduction in East Scotland coincident with periods of daytime low water of spring tides in late October or early November (Figure 6.3) (Watson *et al.*, 2000; Bentley *et al.*, 2001). Similarly, for the ragworm *Nereis virens* it has been shown in the laboratory that the worms possess a circa-annual rhythm of spawning, whereby individuals reproduce within a certain 'window of opportunity' determined by the photoperiodic cycle of annual changes in day length. In a given year, worms that are physiologically ready to spawn do so when the window opens but those that are not at the appropriate stage of maturity during the short period of time when the window opens wait a further full year before the opportunity to spawn presents itself again (Bentley *et al.*, 2001).

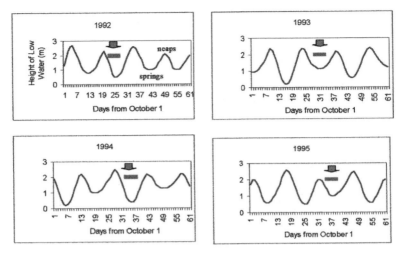

Figure 6.3 Spawning periods of the lugworm *Arenicola marina* at St Andrews, UK, from 1992–95 (after Bentley *et al.*, 2001; data from Watson *et al.*, 2000; by permission of Springer Scientific and Business Media).

If it proves to be a reasonable generalization that annual reproductive cycles of this kind are controlled by a combination of endogenous rhythmicity and a hierarchy of exogenous responses to fine-tuning external variables, it would also be reasonable to reconsider whether the intensity of moonlight can be considered as one of the external variables that cues responsive behaviour or entrains biological clockwork. Certainly, the responses of Atlantic palolo to moonlight (Clark and Hess, 1940) can be more readily understood on the basis of such a model. Also, in *Nereis virens* on British shores, it has been shown that its dawn and dusk foraging behaviour is controlled by extremely low levels of incident light. The worms can detect the equivalent of nautical twilight, even though they are burrow-dwellers living in often turbid water, levels of detection well below that of moonlight in clear water (Bentley *et al.*, 2001). Moreover, as discussed in an earlier chapter, exposure to artificial cycles of moonlight is capable of inducing a circalunar rhythm in the worm *Platynereis dumerilii* (Hauenschild, 1960), the time of cessation of moonlight being the critical cue for establishing the phase of the induced rhythm. If, as seems likely, Samoan palolos are equally sensitive to moonlight and possess an endogenous circalunar clock, then their first exposure to a lunar cycle during the annual reproductive window could set the phase of that clock, so inducing synchronized swarming at the next or subsequent third lunar quarter.

But the story is likely to be even more complex because there are differences in the swarming times of palolos between the various islands of Samoa. Whereas spawning occurs around dawn around the main islands of Savaii and Upoli, it occurs earlier, around midnight, on the island of Tutuila, and just after dusk on Manua, respectively about 70 and 140 km to the east of the main islands. These differences do not appear to be related to differences in tidal timings between the various islands since on Savaii and Upoli spawning occurs during low tide, on Tutuila around high tide and on Manua during the flooding tide. However, since Samoan palolo swarming strengths are now documented by local fisheries biologists, it should, in due course, be possible to formulate hypotheses to explain the timings and pattern of spawning, and to make comparisons between Samoan and West Indian species of palolos. Preliminary tests of such hypotheses might be carried out by comparing spawning strengths and timings with prior weather conditions and moonlight intensities over the reefs, alongside critical experiments carried out using worms that have been cultured in controlled conditions in the laboratory.

*   *   *   *

The swarming of palolo worms is an example of a critical event in the life cycle of an animal that is timed by a presumed circa-annual clock system. This ensures synchronized reproduction and enhances survivorship of the species. Spawning is accurately phased to a particular time of year, phase of moon and time of day as environmental 'windows' open, permitting expression of endogenous rhythmic processes that approximately match the periodicities of the environment. In technical terms the annual swarming rhythm of palolo could be hypothesized to be under the control of a hierarchical sequence of 'gated' rhythms, slightly changing the metaphor from windows to gates. Indeed this kind of hypothesis is probably the most economical to explain a wide range of similar annual clock-related breeding rhythms that have occasionally been referred to in the scientific literature, somewhat pejoratively, as anecdotal. Korringa (1957) documented many such examples, at the same time expressing reluctance to accept the possibility that some annual spawning rhythms might be under direct control of moonlight. His scepticism was based on extensive studies that he had undertaken on the reproductive biology of the oyster *Ostrea edulis*. Daily, from mid-June to late August, for ten years, counts were made of the number of oyster larvae swimming in a unit volume of seawater above oyster beds in Dutch estuaries. Every year there was one great maximum of swarming that occurred between 26 June and 10 July, usually about ten days after the full or new moon that occurred during this period. Spawning, that is the production of eggs and sperm, occurred eight days before swarming, which is defined as the time when swimming larvae were released after spending their first days of development within the parental gill chamber. Therefore, since maximum spring tides above the oyster beds occur two days after full and new moon, it was concluded that the critical phasing event was water pressure associated with those maximum tides. Maximum hydrostatic pressure at the spring high tide was assumed to trigger spawning which occurred two days later than full or new moon, followed by eight days of larval development within the maternal mantle cavity, totalling ten days before swarming. Other workers, too, have sought to characterize external factors as the sole synchronizing agents for the calendar-clock of oysters (Knight-Jones, 1952), the possibility of physiological clockwork phased by lunar cues having been so far excluded from consideration. Korringa's scepticism concerning the possibility that rhythmicity might be phased directly by moonlight also extended to his assessment of the timing mechanism of the summer nuptial dances of the Mediterranean marine worm, *Platynereis dumerilii*, during the first and third lunar quarters. It was only later that Hauenschild (1960) showed unequivocally for the first time that the worms possessed a biological clock of circalunar periodicity and that the clock could be reset by exposure to artificial moonlight (see Chapter 5).

Korringa (1957) also describes several other examples of biological lunar periodicity that may well deserve further study. Very early in the twentieth century anecdotal accounts were reported of moon-related annual swarming by fireworms *Odontosyllis phosphorea*, at Departure Bay in British Columbia and the related *Odontosyllis enopla* around Bermuda. Swarming begins with the emergence into the water column after sunset of luminescent females to which males are attracted, when spawning and egg fertilization occur. Interestingly, Departure Bay fireworms, like Atlantic palolos, swarm at the first and third lunar quarters, whereas Bermuda fireworms, like Samoan palolos, swarm only during the last quarter of the moon. Again there may be survivorship advantages in the release of free-swimming larvae during neap tides but, as with Samoan palolos, the question again arises as to why Bermudan fireworms do not swarm during the first quarter neap tides. Perhaps this again implicates direct influences of the moon, rather than indirect neap tidal influences, in cueing the lunar component of the fireworm biological clock system.

Many other examples of annual rhythms linked to lunar phase are known throughout the animal and plant kingdoms. A particularly spectacular example among fish is the annual spawning frenzy of the grunion fish *Leuresthes tenuis* referred to earlier (Thompson, 1919; Clark, 1925). This small, sardine-like fish lives in the open sea but deposits its eggs in sand at upper levels of beaches in southern California. Spawning occurs particularly in April and May, when males and females ride the waves of rising tides to swim upshore at night. The event is timed during a short interval of one to three days after the times of full and new moon, that is during spring tides. At high tide, at the top of the beach, and partly exposed to the air, the fish dig pits into which the females and males deposit eggs and sperm simultaneously, before rushing seawards again. Maturation of the fertilized eggs to young fish then takes place whilst they remain partially buried in the warm, moist sand that is probably not fully immersed during the following neap tides that occur seven days after the eggs are laid. Remarkably, though, the young fish are ready to swim after 15 days, exactly the time to the next spring tides when they are washed out from the sand and return to sea. Individual fish are known to spawn on successive spring tides, suggesting that the ripening of eggs and sperm is also in phase with the neaps/springs sequence of tides. A model that postulates gated endogenous rhythms of annual, lunar, solar and tidal periodicities could explain the timing of grunion spawning. On such a model, since the fish exploit successive spring tides it is likely that they do not utilize moonlight directly as a cue, but are synchronized indirectly to the moon through the action of tides. That is not to say, however, that fish are unable to detect moonlight. Even the reclusive coelacanth, which occurs at greater depths than grunion, is reputed by Comoros Islands fishermen to be particularly responsive to moonlight and to be more venturesome at the times of full moon (Weinberg, 1999).

A particularly striking, and well-documented, example of an approximately annual breeding rhythm is that of the sooty tern (*Onychoprion fuscatus* = *Sterna fuscata*) (Plate 13). This is a marine bird of tropical oceans which breeds on tropical islands, one sub-species of which nests on the cliffs of Ascension Island situated in mid-Atlantic virtually on the equator. Often called Wideawake Terns by people disturbed by the cacophony of sound around the breeding colonies, nesting occurs at fairly predictable times dependent upon the phases of the moon. Based on observations of the breeding of the terns on Ascension Island from 1941 to 1958, inclusive, Chapin and Wing (1959) formulated what they called 'The Wideawake Calendar'. Using 18 years of observations the authors recorded the day of first arrivals at the nest sites and the dates on which eggs were first reported during each breeding season (Figure 6.4). Initially it had seemed that breeding occurred randomly throughout the

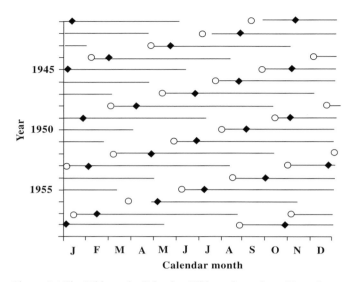

Figure 6.4 The Wideawake Calendar: Wideawake or Sooty Terns *Sterna fuscata* arrive to nest on Ascension Island shortly after every tenth full moon (open circles), the first eggs (diamonds) are laid within a few weeks of arrival and the birds leave after six or seven months in residence (after Menaker, 1976; redrawn from Chapin and Wing, 1959; by permission of University of South Carolina Press).

year, but the calendar shows that in fact the birds returned to the island for breeding around the times of every tenth full moon. There then followed a noisy six weeks or so, subsequently referred to as the 'Night Club' phase, or 'Wideawake Fair', before egg-laying began. Records maintained by the UK Royal Society for the Protection of Birds confirm that the same general pattern of breeding behaviour continues into the twenty-first century. The pattern is a remarkable biological phenomenon and, as such, a seemingly infallible cycle of ten lunar months might generate scepticism of the data. However, from what would be expected based on an understanding of biological variability, the birds occasionally err and return after 11 lunar months. Moreover, in some years breeding fails altogether. Every 12th or 13th full moon would be a closer approximation to the solar year, but natural selection in the terns appears to have favoured adaptation, almost perfectly, to a decimal, not a duodecimal system, of counting months. It is difficult to avoid the conclusion that since they are normally dispersed and breed so close to the equator, sooty terns use every tenth full moon as the cue that triggers their presumed, circa-annual breeding rhythm.

This raises a general question concerning the nature of the environmental zeitgebers that serve as cues for circa-annual rhythms since, in temperate regions, where breeding occurs truly annually, terns and other birds no doubt use seasonally distinct, yearly changes in day length as synchronizing cues. Responsiveness to seasonal changes in day length, or photoperiodic time measurement, is the procedure whereby organisms sense the season of the year by measuring the duration of night or day, which is fundamental in responding appropriately concerning seasonal patterns of growth and reproduction. The phenomenon has been well studied in terrestrial organisms, initially in plants (Thomas and Vince-Prue, 1982) and subsequently particularly in insects (Saunders, 1982), birds and mammals (Follett, 1982; Gwinner, 1989). Quite early it was established that photoperiodic time measurement could be explained as a function of circadian clockwork, rather than necessarily by circa-annual clockwork, of the organisms concerned (Saunders, 1982), a useful model of the process having been postulated by Bünning (1960). The latter author proposed that a plant's circadian cycle consisted of two half-cycles which differed in their sensitivity to light. These constituted a 'photophil', or light-requiring half-cycle, followed by a 'scotophil', or dark-requiring half-cycle. In this way a plant could respond to seasonal changes in day-length by recognizing short days when daylight was restricted to the photophil half-cycle, and long days when daylight extended into the scotophil half-cycle of its clockwork. Such a model for recognition of seasonal changes in day length has potential application in a wide range of terrestrial organisms whether or not they also possess endogenous circa-annual clockwork, but its use seems inadequate for equatorial organisms where annual variations in day length are small. As an explanation of the breeding rhythms of, for example, palolo worms, it seems unlikely. In that species the possession of true circa-annual clockwork phased by moonlight appears to be the most satisfactory model to explain their approximately annual breeding patterns linked ultimately to lunar phase.

The phenomenon of photoperiodic time measurement appears not yet to have been investigated in detail amongst aquatic marine organisms, except in some polychaete worms (Bentley *et al.*, 2001) and in intertidal macrophytic algae (Lüning, 1991; Lüning and Kadel, 1993) from marine intertidal habitats of mid-latitudes where they would be subjected to significant changes in daylight and day length. On British shores it has been shown for the ragworm, *Nereis virens*, that the environmental photoperiodic cycle opens a 'window of opportunity' for the time of reproduction, which is determined approximately by an endogenous circa-annual rhythm of maturation (Olive *et al.*, 2000; Bentley *et al.*, 2001). In addition it has been shown that photoperiodic time measurement serves as the zeitgeber for circa-annual rhythms of growth and reproduction in *Pterygophora californica* and several species of another brown alga, *Laminaria*. For example, Dieck (1991) showed that *Laminaria setchelli* maintained in continuous light/dark conditions of long day and short night exhibited a free-running rhythm of growth and reproduction with classical circa-annual periodicity varying between 11.3 and 17.3 months. In contrast, the rhythm was of precisely 12-month periodicity when the plants were cultured in annual cycles of experimental changes of day length which simulated those of the changing seasons. Moreover, when simulated annual cycles of changing day length were phase-shifted by 3, 6 and 9 months the induced growth and reproductive cycle was rephased accordingly (Figure 6.5) (Dieck, 1991). All of the plants were in a 'winter state' of rest when the experiments began and the control plants grown with no phase change of the imposed annual pattern of light/dark changes showed maximum growth three to four months after artificial day length was increased from its minimum of 8 h. Hapteral growths of the holdfast and reproductive sori also appeared at about the same time. Plants exposed to a three-month (90 degree) phase shift of annual cycle of changing day length at first expressed a growth and reproductive peak to which they were endogenously committed, but their next growth peak was advanced by about three months. In the latter case peak growth and appearance of sori now occurred within a few months after the onset of increasing day length in the phase-shifted annual cycle. Plants introduced into long day lengths, experiencing six- and nine-month phase shifts of 180 and 270 degrees, respectively, showed little sign of growth in the experiments until after they had next experienced short days. There was some evidence of sori production ahead of growth onset in the six-month phase-shift experiment, indicative of endogeneity of reproductive rhythmicity. However, in these two cases, sorus production and, in one case, hapter production, were clearly phase shifted proportional to the artificially advanced time of onset of increasing day length.

Thus the annual growth and reproductive patterns of *Laminaria setchelli* (Dieck, 1991) and other species of the same genus (Schaffelke and Lüning, 1994) appear to be based on persistent circa-annual biorhythms, with clear-cut synchronization by the process of photoperiodic time measurement. It is reasonable, therefore to consider the adaptive advantage of such endogenous time-keeping ability in algae zoned, as they are, in the low intertidal or immediate subtidal regions of coastal seas. In such

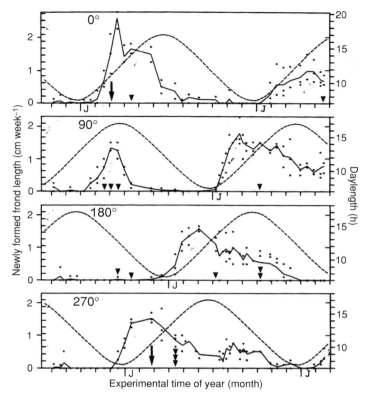

Figure 6.5 Annual growth rhythm of *Laminaria setchellii* (continuous lines) in simulated annual cycles of day length at latitude 54 degrees (dotted line), at 5°C. The annual cycle of day length is phase-advanced from the control by 90, 180 and 270 degrees, giving simulated three, six and nine months advances, respectively, of the imposed annual cycles of day length change. (Dots: individual plant growth rates; arrows: first visible sori; arrowheads: initiation of new ring of haptera in individual plants.) (After Dieck, 1991; by permission of Wiley-Blackwell.)

localities it would not be surprising if the seasonal changes in day length that synchronize algal circa-annual rhythms might equally provide accurate exogenous cyclical clues for growth without the plants having to incur the physiological cost of adaptively acquiring innate circa-annual clockwork. However, Dieck (1991) proposed that it may be advantageous to have evolved endogenous circa-annual rhythmicity in view of possible obscuring effects upon the day length signal of cloudy weather, ocean turbidity and tidal action. Also, in *Laminaria digitata* and *L. hyperborea* it has been shown that the periodicity of the free-running growth rhythm is shorter than annual. This inclined Schaffelke and Lüning (1994) to suggest that such endogenous rhythmicity, coupled with an ability to respond quickly to environmental signals, characterizes the *Laminaria* species in question as 'season anticipators' (see also Kain, 1989), which become largely independent of ambient irregularities. In addition, as discussed in Chapter 2, it seems likely that there is considerable adaptive advantage in circa-annual rhythmicity if it triggers preparatory physiological processes which of necessity precede growth and reproduction. However, as Dieck (1991) points out for algae, the adaptive value of circa-annual rhythms in contrast to exogenous photoperiodic control systems will only be fully understood with better understanding of the zeitgeber action of light and the role of phytochrome systems, as discussed for terrestrial plants by Vince-Prue (1982) (see Chapter 10).

* * * *

Motivation for understanding the synchronization and timing of spawning in marine organisms is not simply attributable to scientific curiosity. Such information can be crucial for fisheries biologists charged with providing advice for the sustainable management of commercially important species of fish and shellfish. For many commercially fished species, long runs of data concerning spawning dates are available, some dating back to the late nineteenth century (Cushing, 1969). Based on fisheries data records for southern North Sea plaice *Pleuronectes platessa*, Norwegian herring *Clupea harengus*, Pacific sockeye salmon *Oncorhynchus nerka* and Arctic cod *Gadus morhua*, each fish was shown to have a characteristic spawning period of a week or two only each year. Moreover, for three of the species, the times of potential spawning days were constant from year to year. Arctic cod alone showed a slight delay in spawning over time, but even that by only about eight days in 70 years (Cushing, 1969). A particular concern of fisheries scientists is whether or not the annual spawning events coincide with springtime production of phytoplankton upon which newly spawned fish larvae feed. If they coincide, it heralds a successful year class of fish; if they do not, then the stock may survive at only low levels during a poor year, hopefully to recover in a later year when spawning and phytoplankton production again coincide. The implication of this match/mismatch explanation of the variability of fish and shellfish population densities is that spawning is not triggered by a particular phase of the annual cycle of the production of phytoplankton, as was once thought to be the case. Cushing (1969) concluded that since spawning times are linked to plankton production cycles in only an indirect manner, it is to the advantage of a fish species if it spawns at a relatively fixed date in their calendar. In this way a fish population has the best chance of profiting from the variability in timing of the plankton production cycle (Cushing, 1969).

From the point of view of the chronobiologist, the repeatability of the timing of spawning cycles is of interest in suggesting that the four species of fish, and some shellfish that have been studied, possess circa-annual biological clocks that control spawning. Direct experimental evidence of the occurrence of such long-period clocks is difficult to obtain, due to the difficulties of maintaining organisms in constant conditions for very long periods of time. However, new opportunities for such studies have arisen from recent developments in the commercial cultivation of fishes. For example, the freshwater rainbow trout, *Oncorhynchus mykiss*, has been maintained in a constant schedule of 6-h light and 18-h dark, at constant temperature and constant feeding rate, for more than four years without apparent ill-effect. Under such conditions, without any clues as to the passage of the seasons, the fish showed free-running circa-annual rhythms of maturation of the gonads which were self-sustaining for three 'years' (Dustan and Bromage, 1991). In marine fishes the endogenous physiological clockwork necessary to control annual spawning cycles can be envisaged as having evolved against the annual periodicity of events in the ocean. In general, seasonal changes in day length and temperature are less well defined in the sea than on land, even in temperate localities, and therefore individually may be only minor synchronizers of fishes' annual biological clocks. However, it is not unreasonable to postulate that circa-annual spawning rhythms of higher latitude fish and shellfish have evolved against seasonal changes in day length and temperature, together with the averaged seasonal timing of plankton production over evolutionary time, even though the timing of plankton production may be more variable than spawning times. Future studies may yet find evidence of circa-annual endogeneity in plankton production cycles. However, on present evidence, annual outbursts of plankton production in spring and autumn in temperate seas can perhaps be most conveniently regarded as exogenously generated events (Naylor, 2005). They probably depend most heavily upon nutrient availability in response to environmental temperature and weather conditions, particularly wind strength and direction, that influence stirring of the sea. Indeed, the broad pattern of seasonality of temperate seas relies heavily upon enrichment by nutrients stirred up from the depths by equinoctial gales and tides during spring and autumn.

However, equally important in our understanding of the calendar-clocks of fish is the comparison that can be made between the seasonal spawning times of plaice, cod, herring and sockeye salmon as compared with those of the Californian sardine and the Pacific tuna, the last two of which spawn at very variable times of year. The first group of four are all temperate or high latitude species, occurring in localities where seasonal changes in day length and temperature are apparent and where cyclonic weather conditions generate ocean stirring, particularly during equinoctial tides of spring and autumn. The timings of all these factors, averaged over evolutionary timescales, can be considered to have contributed to the process of natural selection for endogenously controlled, marked seasonality of breeding of fish and many invertebrates in high latitudes. By contrast, the Californian sardine and Pacific tuna which live at lower latitudes, where seasonality is much less pronounced, breed at more variable times of the year (Cushing, 1969). In such localities it seems likely that there has been no evolutionary advantage in the acquisition of sharply defined annual breeding times such as occur in higher latitude species (Naylor, 2005). Indeed, though they initially appear to be precise, spawning times of tropical palolo are less sharply defined in an annual sense than those of high latitude fish, in that they may occur at almost any time during the two months of October and November. Precision in any one year in palolo swarming is apparently achieved by response to lunar phase in the absence of the major seasonal variables that occur in higher latitudes. The tropics and subtropics are regions of anticyclonic, rather than cyclonic, weather conditions; moonlight there would be expected to be a more reliable synchronizer than it is in higher latitudes. Perhaps it is not surprising that many, though not all, examples of lunar related rhythms in marine animals occur in species that live nearer the equator than the poles.

*   *   *   *

In this chapter it has been shown that, just as there are circadian, circatidal, circasemilunar and circalunar clocks in marine animals there are probably also circa-annual clocks and that all these clock types may interact to determine the pattern of animal calendars, though hard evidence for endogenous circa-annual rhythmicity is meagre. Many species worldwide are known to breed annually, often over lengthy periods during a year. For example, the annual breeding season of loggerhead and green turtles on Queensland beaches in Australia extends from December to March. Adults which have migrated inshore emerge onto the beaches at dusk and move upshore to lay their eggs, before returning to the sea immediately, to be followed by the hatchlings about six weeks later. However, the evidence for circa-annual clock control of such behaviour is largely circumstantial. Nevertheless, some evidence is available from experiments involving marine worms, fish and algae that it is possible to maintain in culture throughout the year. Further studies on annual breeding calendars of marine organisms seem likely to be required in the future as more species are considered for commercial culture. The use of controlled regimes of light and dark in commercial aquaculture practices needs to be fully evaluated against not only circatidal, circadian and circalunar biorhythms in the animals concerned, but also against the likely existence of internal clocks of circa-annual periodicity (Berrill *et al.*, 2003).

7

Plankton vertical migration rhythms

Almost throughout the oceans and seas . . . as night approaches very great quantities of animals migrate upwards from the deeper, dark waters towards the surface layers, where they spend the night until at dawn they descend again to their daytime residence depths

Alan Longhurst, 1976

The first recorded observations of the phenomenon of vertical migration by aquatic animals were published early in the nineteenth century, by naturalists studying the plankton of freshwater lakes (Cushing, 1951). By the end of that century pioneering biological oceanographers had also observed that greater numbers of plankton animals were caught at the sea surface at night than by day. It was then that the hypothesis was first formulated that marine zooplankton undertook daily rhythms of vertical migration. Later, oceanographers began to utilize specially designed tow nets to capture plankton at different depths throughout the day and night and so test the hypothesis. There were initial difficulties in interpreting early findings since the first tow nets to be used were open when they were lowered into the water and remained open until they were hauled on deck (Murray, 1885). So, although the nets were towed at a particular depth for a known interval of time, they inevitably also caught plankton on the way down and on the way up, clearly presenting problems when trying to determine the precise depth at which particular animals were collected. Those problems then stimulated, at the beginning of the twentieth century, the invention of opening and closing nets that could be kept closed until lowered to a selected depth, at which they were opened by a messenger weight that was slid down the towing wire (Michael, 1911). Then, before the net was raised, another messenger weight was sent down the wire to activate an additional mechanism to close the net, a technique that was subsequently replaced by acoustic opening and closing mechanisms. These breakthroughs made it possible to lower a closed net from a research ship to a predetermined sampling depth, open it and sample plankton at that depth, and then close the net again before hauling it to the surface. In this way samples could be obtained from precisely determined depths at various times of day and night, permitting confirmation of the daily vertical migration hypothesis, as a basis for the establishment of the theory of plankton migration. Studies off California (Esterley, 1911), in the western English Channel (Russell, 1925, 1927) and in the Antarctic Ocean (Hardy and Gunther, 1935; Frazer, 1937) emphasized the global significance of the phenomenon such that by the early part of the twentieth century it became axiomatic that, generally in the ocean, plankton animals exhibit vertical migration rhythms of daily, or diel, periodicity (DVM). The phenomenon was established as a paradigm, underlying an understanding of the relationships between plankton and commercially important fish exploited by humans.

Later, even more precise plankton sampling was undertaken by pumping seawater from predetermined depths and filtering off the plankton it contained aboard ship. Then highly sophisticated gadgetry was developed in the form of a continuous recorder, the Longhurst–Hardy Plankton Recorder, which collected plankton on a moving scroll of fine mesh netting (Longhurst, 1976). The mesh unwound from one spindle and re-wound on to another, capturing plankton as it did so over the duration of the sampling transect, as the instrument was towed behind a research vessel. Back in the laboratory the preserved samples were unwound and analysed to provide very precise information concerning the geographical position and depth of capture of the organisms caught in the mesh. Even more sophisticated developments of the original recorder are now routinely deployed from ships of passage in the north Atlantic. These modern sampling devices provide very precise information about the vertical migrations of plankton. In addition, they permit the development of synoptic geographical records of plankton species at particular depths at various times of the year as a basis for improved understanding of the interactions between ocean plankton and fluctuations in commercial fishery yields and in assessing the impact of human-induced changes in the ocean against the background of natural variability.

Since the initial characterization of DVM in the early part of the twentieth century, research interest in the phenomenon by marine biologists and oceanographers continued unabated except for a few years in the early 1940s when the hazards of naval warfare in World War Two placed serious restrictions on the kind of basic research that could be carried out in many of the world's seas. By chance, however, the study of marine plankton was not neglected during that time; indeed the daily migration of plankton was ' re-discovered' serendipitously by naval researchers (Longhurst, 1976). The development of acoustic depth indicators and acoustic submarine hunting devices (sonar) during early stages of the war quickly led to the discovery of echoes from a 'deep scattering layer' almost everywhere in the world's oceans at a depth of 100 to 400 m or more, depending upon the time of day. The strong echoes from this layer were at first sometimes confused with echoes from the sea bed itself, until it was also realized that part of the deep scattering layer appeared to rise towards the surface at dusk and to descend again at dawn. It was several years before it was confirmed that the layer was indeed the reflection of sound scattered by a layer of animals, subsequent samples of which included coelenterate medusae, crustaceans, cephalopods and small fish (Hersey and Moore, 1948). Eventually, too, the findings were confirmed by direct observation from submersibles (Barham, 1963). Importantly, recordings of the deep scattering layer not only confirmed the occurrence of vertical migration in many mid-water species, but also made fundamental findings concerning the size and scale of the layers of animals concerned. The layers may be less than a few tens of metres in thickness, sometimes less than 10 m, and may be traced horizontally for thousands of kilometres (Longhurst, 1976), with significant implications for commercially important fisheries in the open ocean.

Some studies of the phenomenon of DVM have also addressed questions concerning possible interactions between vertically migrating zooplankton and the bottom-living communities of animals living in shallow seas. For example, Barr (1970) showed that pink shrimp *Pandalus borealis* in Alaskan waters, a species usually caught in bottom trawls, quite clearly fed on zooplankton. Moreover, they did so during the early part of the night when zooplankton species would be expected to be migrating upwards in the water column. Bottom trawling for shrimp clearly showed greater availability during the day than at night, suggesting that they, like the zooplankton, also moved upwards into the water column at night. This observation was confirmed by around-the-clock sampling of shrimp using baited creels suspended at various depths in water of 90 m depth. Catches showed that growing individuals, in particular, left the bottom, evidently to feed on truly planktonic species from sunset to dawn. Similar observations of nightly upward migrations by another supposedly bottom-living species of shrimp, *Dichelopandalus bonnieri* (Plate 14), in the Irish Sea (Al-Adhub and Naylor, 1977), confirm the behavioural link-ages between planktonic and benthic communities, as do observations of migrating planktonic communities reaching the sea bed during vertical migration. Typical planktonic species such as chaetognaths *Sagitta elegans*, euphausiids *Thysanoessa raschi* and calanoid copepods *Metridia longa* have been recorded reaching the bottom during DVM in a Norwegian fjord of 180 m depth, thus contributing to the epibenthic community just above the muddy sea bed (Hopkins and Gulliksen, 1978).

Such studies of bentho-pelagic coupling were originally of interest concerning the behaviour of bottom-living fish and invertebrates of com-mercial importance where capture is by bottom trawling and population assessment may be subject to daily variations in abundance and size com-position. As Barr (1970) pointed out, an intriguing potential application of such understanding concerning the fishery for pink shrimp was to sug-gest that commercial catches should be restricted to night-time trawling. Night catches would be lower than those taken by day but the average size of the individual caught would be higher due to the migration off the bottom of the growing individuals. This procedure would produce higher quality yields than daytime catches, whilst preserving the smaller mem-bers of the population which, when caught, are often discarded dead. But studies of bentho-pelagic coupling have become increasingly impor-tant in recent years concerning, for example, climate change, requiring a fuller understanding of the processes involved in the transfer of atmo-spheric carbon above the ocean, via phytoplankton and zooplankton food chains, to bottom sediments which act as carbon sinks.

*   *   *   *

Until the latter part of the twentieth century much of the biological work on DVM of marine plankton was concerned with the movements of indi-vidual animal species, with only intermittent sampling. That approach lacked continuity and ignored possible species interactions, prompt-ing more comprehensive studies of entire plankton communities over extended periods of time. One such study carried out aboard the Royal Research Ship *Discovery* (Roe, 1984), involved sampling by day and night for 14 days always within a 9.3-km radius of a point in the middle of the north east Atlantic. Over successive 48-h periods, sampling was car-ried out every other hour at depths of 100, 250, 450 and 600 m, using acoustically opening and closing nets so that there could be no risk of contamination of samples as the nets were lowered and raised from the appropriate sampling depth. The main aim was to determine the differ-ent times when vertically migrating plankton migrated through the four sampling depths. It was found that the proportions of animal groups and

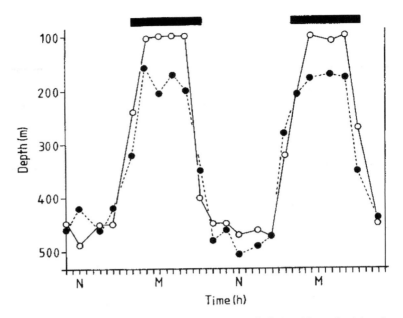

Figure 7.1 Mean depths (m) of adults (open circles) and juveniles (closed circles) of the deep-ocean planktonic prawn *Systellaspis debilis* during 48 h of continuous sampling at a location in the north east Atlantic. N, noon; M, midnight (redrawn from Roe, 1984).

species varied continuously on a daily basis at all depths and particularly so at 100 and 250 m. The changes were clearly due to vertical migrations. Some species occupied upper levels in the sea by day and migrated to the shallowest depths by night, while others living deepest by day migrated only to mid-depths at night. A particularly abundant prawn-like crustacean, *Systellaspis debilis*, was centred between 400 and 500-m depth by day but swam upwards to a depth of 250 m by sunset and reached 100-m depth early in the night (Figure 7.1). Downward movement started well before sunrise, reaching 250 m by dawn and 450 m by midday. Daytime migration of some individuals extended down to 600 m by day, from where upward migration began again in the early afternoon (Roe, 1984).

One of the first explanations of the biological advantage of this kind of daily vertical migration in planktonic animals was formulated late in the nineteenth century (see Cushing, 1951; Longhurst, 1976). It stated simply that herbivorous planktonic animals preferred darkness, and that vertical migration was a feeding migration because the plant plankton upon which they feed lies in the surface layers. At the scale of movements exhibited by zooplankton the phytoplankton layer remains relatively static. Some phytoplankton groups such as the flagellates can migrate vertically over short distances by flagellar action, but major groups such as diatoms can sink or rise only slowly, by changes in buoyancy (Steele, 1976). On this interpretation, carnivorous feeders would also swim upwards at night to feed on near-surface aggregations of their herbivorous prey, as indeed basking sharks feeding on zooplankton have been shown to do (Shepard *et al.*, 2006). Carnivorous mid-water plankton species are, however, not simply following particular prey. A number of crustaceans that have been studied all appeared to feed during their entire vertical migrations, and their diets were different when caught at opposite ends of their migrations (Roe, 1984). A converse mechanism put forward is that plankton species descend from surface waters by day because near the surface they would be vulnerable to predation by visually feeding fish and marine birds (Longhurst, 1976). Indeed, Hays (1995) showed that in the North Sea, marked changes occurred between 1960 and 1990 in the size of the herring *Clupea harengus* stock which could be correlated with changes in the vertical migration behaviour of its important prey species, the copepod *Calanus finmarchicus*. In the years when herring were most abundant a smaller than average proportion of *Calanus* occurred in surface waters by day, presumed to be related to an increased risk of mortality. As Vinogradov (1968) wrote 'high concentrations of zooplankton in the surface layers facilitate the predatory activities of fishes [so that] by sinking and dispersing the zooplankton largely escapes annihilation'. However, the two suggestions are not mutually exclusive, nor do they exclude another suggestion relating to the need to conserve energy. For example, observations from manned submersibles show that some fish and crustaceans hang motionless in the water during daytime resting phases at depth (Herring and Roe, 1988). Also, as will be discussed in Chapter 9, vertical migration may enable some zooplankton species to exploit water currents at different levels in the sea as the basis for lateral transport over extensive geographical distances. In any case, the search for a general hypothesis to explain the evolutionary advantages of daily vertical migration became somewhat futile when later findings questioned whether animal plankton always swam upwards at night. There were clues to this possibility in echolocation studies which suggested that only part of the deep scattering layer moved upwards at night (Longhurst, 1976) and then later findings clearly demonstrated anomalous downwards swimming behaviour at night by

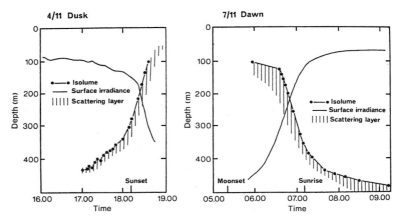

Figure 7.2 Dusk and dawn correspondence between isolume, surface irradiance and a scattering layer near the Canary Islands (after Longhurst, 1976; redrawn from Boden and Kampa, 1967; by permission of Wiley-Blackwell).

the appropriately named copepod crustacean *Anomalocera*, presumed to be an adaptation to avoid nocturnal predators (Tester *et al.*, 2004). Furthermore, it has been shown that in some mixed populations of animal plankton species, not all of them migrated upwards at night and those that did so often migrated over different distances according to their stage of development and the density of their predators (Neil, 1992; Irigoien *et al.*, 2004). Nevertheless, vast quantities of animal plankton do undertake upwards vertical migrations at night and the question remains as to how that is achieved.

* * * *

It is apparent that in the open ocean environment there are easily identified vertical gradients of temperature, water chemistry, density, pressure and light, and the last of these is probably the single major environmental factor that most influences vertical migration by animal plankton. In passing downwards, even through transparent oceanic waters, sunlight is quickly scattered and the red part of the spectrum is absorbed, such that at 200-m depth an animal's surroundings are uniformly blue-green in colour. In some oceanic locations, even down to depths of 900 m or so there is an adequate light signal to support the conclusion that light is an important cue for vertical migration (Forward *et al.*, 1984; Forward, 1988; Herring and Roe, 1988).

　　Early indications were that animal plankton species follow a preferred light intensity zone as it rises and falls throughout the day and night, giving rise to the so-called 'isolume hypothesis' (Figure 7.2) (Rose,

1925; Russell, 1927; Boden and Kampa, 1967). The hypothesis came to be attractive since it had been shown that Arctic Ocean zooplankton ceased to undertake daily vertical migration in summer during virtually continuous 24-h daylight (Bogorov, 1946), notwithstanding contrary evidence that vertical migration did seem to persist in summer in Tromso Sound, Norway (Marshall and Orr, 1955) and around Spitzbergen (Digby, 1961). The hypothesis was also sustained by opportunistic oceanographers who recorded short-term rises in the deep scattering layer during episodes of daytime darkness induced by a total eclipse of the sun (Skud, 1967). Unfortunately, this attractively simple suggestion cannot be the whole story (Roe, 1983; Herring and Roe, 1988). It implies that layers of different zooplankton species, arranged like the steps of a ladder, move up and down throughout a 24-h period, maintaining their relative positions as they do so. However, to follow an isolume a population of planktonic migrants would need to be confined within a very narrow depth-layer, which is narrower than many observed in practice, and would need to be able to swim faster than has been observed for many species in order to keep pace with isolume depth changes at dusk and dawn (Roe, 1984). Also, having migrated upwards at dusk, many species temporarily sink again around midnight, falling below a presumed optimum light zone, only to swim upwards again before the dawn sinking. Some examples of this phenomenon may be attributed to the effect of moonlight at the surface, but it seems unlikely that all cases of midnight sinking can be explained in that way, in view of lunar monthly variations in the intensity of moonlight (Herring and Roe, 1988). Strikingly, too, several species are known to overtake others on the way up and on the way down again (Roe, 1974), clearly showing no preference for an optimal light zone as they do so. Such changes in the relative vertical positions of different components of the zooplankton strongly militate against the so-called 'ladder of migrations' that the isolume hypothesis implies. This interpretation is supported by detailed studies of the depth distributions of two species of copepods in the Irish Sea by Lee and Williamson (1975). In those studies the day-depths of *Pseudocalanus elongatus* and *Microcalanus pusillus* were found to vary considerably at any one station, the former showing diurnal migrations on some but not all occasions. Some of the vertical distributions of both species, by day and over the 24-h cycle, could be correlated with physical or chemical characteristics of the water layer in which they were found, water chemistry being of particular significance. Moreover, detailed studies of mixed populations of species have demonstrated that not all species show vertical migration, and those that do sometimes show differences according to their stage of development and risk of predation by larger animals. Again, in Antarctic krill *Euphausia superba*, it has been shown that dense swarming occurred during daytime, followed, not by migration, but by dispersal at night (Everson, 1982). Similarly, in a study of vertical distribution of copepod crustaceans at four localities in the Arctic Sea (Daase *et al.*, 2008) it was observed that diel vertical migration occurred in only the older stages of development of *Metridia longa*; younger stages remained in deeper water during both day and night. Some stages of *Calanus* spp. were also found to be non-migratory, whilst others did migrate, the latter suffering lower population mortalities than the non-migrating individuals. Also, many zooplankton species are known to show variation in the distance and direction of their migrations, some even migrating downwards at night and upwards by day, the reverse of the pattern characterized by the plankton paradigm (Irigoien *et al.*, 2004). Furthermore, it has been shown that tagged basking sharks sometimes show daily vertical migrations in reverse of their normal pattern by swimming deeper at night, presumed to be in response to unusual downward swimming at that time by their zooplankton prey (Shepard *et al.*, 2006).

* * * *

Though twenty-first century studies have questioned the universal nature of daily vertical migration by ocean plankton, thus challenging the plankton paradigm that was based on earlier observations, the extent of the phenomenon still raises important questions as to how it is achieved. Herring and Roe (1988) concluded that daily changes in light intensity, whatever the nature of light at depth, probably comprise the single major environmental influence controlling vertical migration behaviour of most zooplankton. Indeed there is direct experimental evidence that light affects the daily locomotor activity cycles of adult and larval shallow water crustaceans (Rodriguez and Naylor, 1972; Naylor, 1982; De Coursey, 1983; Forward *et al.*, 1984). Even so, since the animals are behaving as living clocks in the open ocean it is also relevant to ask whether, like animals living elsewhere in the earth's biosphere, planktonic animals possess endogenous biological clock mechanisms that permit them to do so. Is their behaviour such that they would migrate up and down, even in continuous darkness or light, or are they simply driven by changes of light intensity during the day/night cycle? In seeking answers to those questions the investigator is immediately confronted with the problem of carrying out laboratory experiments on deep sea plankton. Many of the animals are fragile and often damaged during collection, and are also difficult to maintain in aquaria for extended lengths of time, even aboard the research ships from which they were collected. Accordingly

many of the studies of biorhythms in zooplankton have so far been concerned with hardy coastal plankton species that survive well in laboratory
conditions.

Probably the first indication that vertical migration rhythms in
planktonic animals might be governed by an endogenous rhythm of
swimming was reported by Esterly (1917) in a paper under the delightful
title of 'The occurrence of a rhythm in the geotropism of two species
of plankton copepods when certain recurring conditions are absent'.
The emphasis of the study was on internally controlled daily changes in
responses to gravity in the copepod *Acartia*, suggesting for two species
of the genus that positive or negative responses to a constant external
stimulus were triggered at different times of day by an internal timing mechanism, and similar findings were reported by Hardy and Paton
(1947) for the related copepod *Calanus*. Also, along the same lines, Moore
and Corwen (1956) studying the vertical distribution of siphonophores
off Florida, found it necessary to postulate a daily rhythm of responsiveness to light, temperature and depth (pressure) in order to account for
their results. Again, Singarajah *et al.* (1967) found an endogenous rhythm
in the planktonic nauplius larvae of the shore barnacles *Elminius modestus*
and *Balanus balanoides*, which were photonegative at midday and pho-
topositive in the afternoon, even in constant conditions, consistent with
their pattern of swimming downwards by day and upwards at night.

Perhaps the first, most clear-cut demonstration that a circadian
rhythm of swimming activity is involved in the daily vertical migration behaviour of plankton animals is that published by Harris (1963).
The marine copepod *Calanus* and the freshwater cladoceran *Daphnia* contained in water-filled, vertically arranged glass tubes were observed to
swim upwards at 'night' and sink by 'day', even when kept in constant
conditions in the laboratory. Subsequently a few hardy crustacean species
from inshore plankton hauls have been found to survive well and to be
amenable to long-term experiments in the laboratory. For example, the
amphipod *Nototropis* and some peltidiad copepods exhibited marked daily
vertical migration behaviour for two days in a light/dark cycle in the laboratory and continued afterwards with such behaviour, though in a less
clearly distinct pattern, for four days in constant dim light (Figure 7.3)
(Enright, 1975). Thus, though for example Herman (1963) was unable to
demonstrate an endogenous component in the daily vertical migration
rhythm of the opossum shrimp *Neomysis americana*, there are nevertheless
sufficient examples that support the suggestion that responses to light are
only part of the zooplankton vertical migration story; endogenous timing processes can also play a significant role (Enright and Hamner, 1967;

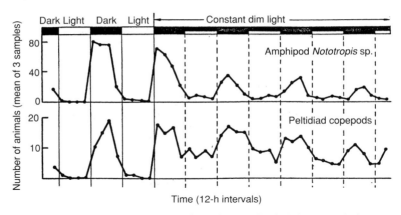

Figure 7.3 Vertical migration of zooplankton in the laboratory during
2 days in light/dark cycles, followed by 3 days in constant dim light.
Numbers of animals on the vertical scale are averages at the surface every
4 h of triplicate samples. Vertical lines indicate 12-h intervals of actual and
expected light and dark. The entraining light/dark cycle was shifted by 6 h
from outside conditions (after Enright, 1975; by permission of
Wiley-Blackwell).

Enright, 1975, 1977; Longhurst, 1976). Among unicellular phytoplankton organisms, too, the luminescent microalga *Lingulodinium* (*Gonyaulax*)
exhibits a truly circadian rhythm of periodic aggregation when kept in
constant conditions in laboratory cultures. The behaviour is considered to
be a type of behaviour which parallels vertical migration of zooplankton.
In the laboratory, peak periods of aggregation occur initially in the early
afternoon, persisting endogenously for several weeks in constant white
light and free-running at a periodicity of slightly less than 24 h. Coin-
cidentally, dispersal of the microalgae occurs at night when vertically
migrating animal grazers of phytoplankton would be arriving in surface
waters. It seems reasonable to assume, therefore, that, in nature, periodical aggregation by microalgae is related to the requirement for photosynthesis, which would be most efficient at the sea surface during daytime
and that dispersal at night reduces grazing pressure by small, herbivorous, vertically migrating microzooplankton (Roenneberg and Morse,
1993).

How generally it turns out that ocean plankton species not only
behave as living clocks, but *are* living clocks, remains to be seen. However
the occurrence of circadian clocks, synchronized by light/dark cycles and
occasionally overridden by environmental perturbations, such as prolonged polar summers and eclipses of the sun, offers a comprehensive

basis for our understanding of the phenomenon of daily vertical migration by many zooplankton species in the open ocean, whether or not that phenomenon can still be regarded as a paradigm for marine biologists and oceanographers.

* * * *

In fact, it turns out that the early generalization that zooplankton, in particular, exhibit nightly migration to surface waters has been challenged by other findings too. As will be discussed in the next chapter, it has been known for some time that a few zooplankton species living in tidal estuaries exhibit not daily, but tidal rhythms of vertical migration, again even when kept in constant conditions in the laboratory. Such observations generated the question as to whether circatidal rhythmicity in zooplankton species was confined to the special case of animals in estuaries, or whether it occurred in marine species too. Are there any open sea plankton species that further challenge the plankton paradigm by exhibiting tidal rather than daily vertical migration? This was a question addressed by Zeng and Naylor (1996a) in relation to the larvae of the common shore crab *Carcinus maenas*. As seen in earlier chapters, this species of crab exhibits clear circatidal rhythms that help to control its intertidal migrations as adults, and the ability of the crabs to behave as living clocks in this way appears to be inherited. It is also the case that the zoea larvae of this crab are hatched and released into the sea at the times of high tide, to be dispersed widely in coastal seas on ebbing tides (Zeng and Naylor, 1996b, 1997). Offshore they form an important component of the coastal, open sea plankton community where they feed on single-celled plant plankton and are fed on by other zooplankton species and by fish. Over a period of a few weeks, the larvae undergo development through four zoea stages before finally moulting to the megalopa (Plate 15). The megalopa larva is an intermediate stage between the zoea and first crab stage, and is the stage at which recruitment back to the shore takes place. In late spring, the time of recruitment on British shores, megalopa larvae of *Carcinus maenas* can often be seen in vast numbers at the water's edge at the high tide level. If swarming occurs in a suitable pebbled location in the upper reaches of a rocky shore, conditions are optimal for the megalopae to then settle out from the plankton and metamorphose into the first crab stage.

As a first step to find out if zoea larvae collected from the open sea exhibited vertical swimming rhythms of tidal periodicity, newly released shore crab larvae were collected from the sea off the coast of North Wales, UK, using plankton nets streamed in tidal flows past a jetty. In the laboratory, in batches of several hundreds, the larvae were then placed in a transparent plastic tower filled with seawater through which clean seawater was dripped slowly to keep the larvae healthy, excess water overflowing through a fine mesh to prevent escapes. The apparatus was placed in a darkened room at constant sea temperature and the swimming behaviour of the larvae was recorded electronically without any human disturbance in the room. Infra-red light emitters and highly sensitive photo-receivers were placed across the top and bottom of the tower in such a way that infra-red beam interruptions by swimming larvae could be logged remotely by computer (Zeng and Naylor, 1996a). During the first 60 h of such experiments, the larvae quite clearly swam most abundantly near the top of the plankton tower at and just after the times of high tide in the sea outside the laboratory and swam most abundantly near the bottom of the tower between those times (Figure 7.4). The young zoea larvae of the shore crab were clearly behaving as living tidal clocks, timed by their own internal biological clockwork. This was a clear-cut, early example, of a circatidal rhythm of vertical migration in an open sea planktonic animal. In the sea, such behaviour would place the larvae near the surface in offshore-ebbing tidal currents, and near the bottom during flood tides where frictional effects slow down onshore water movement. Downwards swimming during rising tides would take the larvae away from surface waters that flood rapidly onshore. Therefore, for the early zoea larvae of shore-living crabs this behaviour would enhance their chances of dispersal into food-rich waters offshore and reduce the risk of them being stranded prematurely between tidemarks.

Further experiments were then carried out to determine how the timing of the larval circatidal rhythm was set and whether the capability to express such a precisely timed swimming rhythm was learned during development or was inherited. Firstly, berried female crabs carrying eggs were brought into the laboratory and maintained in tanks with circulating seawater away from the influence of tides. Eventually these crabs released zoea larvae, the swimming patterns of which were then tested in the plankton tower. Surprisingly they too exhibited vertical migration rhythms of approximately tidal, that is, about 12.4-h, periodicity, even though they had never experienced such an environmental periodicity before, except as eggs when attached to the parent female before capture (Zeng and Naylor, 1996b). If the times of upwards swimming had coincided with the times of ebbing tides on the beach outside the laboratory, it would have been necessary to consider whether the larvae could in some way have detected the rise and fall of tides even when swimming in the plankton tower. In that case the original hypothesis that they possessed internal biological clockwork might be wrong. However,

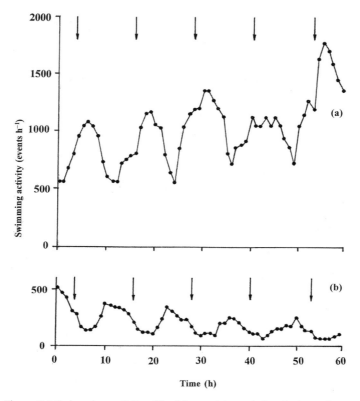

Figure 7.4 Swimming activity of freshly caught, newly hatched zoea larvae of the shore crab *Carcinus maenas* in a plankton tower in constant conditions in the laboratory. Larvae swam in greatest numbers near the top of the tower (A) at times of expected high tide (arrows) and near the bottom (B) at times of expected low tide (after Zeng and Naylor, 1996b; by permission of Inter-Research).

further analysis of the swimming rhythms of laboratory-hatched larvae confirmed that the original hypothesis was correct. Though the rhythmic pattern of swimming of laboratory-hatched larvae was similar to that of larvae freshly caught in the sea, the timing of their episodes of upwards swimming bore no relation to the timing of tides on the adjacent coasts. As mentioned, the eggs from which the laboratory larvae were hatched would have been exposed to tides whilst they were being carried by the berried females prior to capture, so it is conceivable that some environmental factor could have set the timing of larval clocks during the course of egg development. However, circatidal swimming rhythms were also recorded in larvae hatched from females that produced eggs when kept

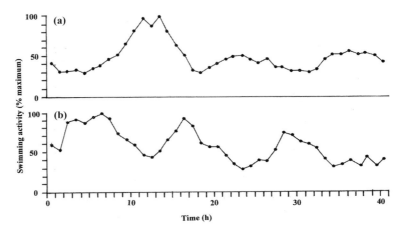

Figure 7.5 Swimming activity of newly hatched zoea larvae of the shore crab *Carcinus maenas*: (a) near the top; (b) near the bottom of a plankton tower maintained in constant conditions in the laboratory. Larvae were obtained from a female crab which acquired eggs after 38 days and released larvae after a further 107 days, entirely when held in the laboratory in the absence of tides. The periodicity of upwards swimming is approximately tidal (12.4 h) despite lack of previous experience of tides (redrawn from Zeng and Naylor, 1996c).

in the laboratory, many weeks after they were captured (Figure 7.5) (Zeng and Naylor, 1996c). On this evidence, it seemed unlikely that the ability to exhibit circatidal swimming rhythms was acquired during the process of egg development and that inheritance of tidal time-keeping ability was the most likely explanation of the tidal swimming behaviour of *Carcinus* larvae.

Further investigation then revealed that the timing of circatidal swimming rhythms in the early stage zoea larvae of *Carcinus maenas* was linked to the time when they were hatched (Zeng and Naylor, 1996b). When a berried female releases her eggs she lowers her abdomen, beneath which the eggs are clustered and attached to swimmerets. At the same time the swimmerets are moved in such a way as to shake the hatching larvae free of the egg membranes. Most females hatch their larvae at the times of night high tides, from which it can be concluded that the phase of the vertical swimming rhythms of larval shore crabs is set by the hatching process. The circatidal swimming rhythm is set in motion with a burst of swimming immediately the larvae are released (Zeng and Naylor, 1996b). These initial bursts of tidally phased swimming would take recently hatched, zoea larvae into surface waters during an ebbing tide, immediately facilitating the process of dispersal offshore. Indeed, further confirmation of this interpretation came about when it was found that eggs, detached from female crabs and kept in circulating seawater in the laboratory until they were ready to hatch, could be induced to produce larvae that behaved as if they had been hatched from a female (Zeng and Naylor, 1996c). All that was required to achieve this was to take mature, detached eggs and gently stir them in seawater for 2 h every 12 h or so for about 48 h. By this means a perfectly phased swimming rhythm of circatidal periodicity could be induced in the recently hatched zoea larvae, upwards swimming occurring immediately after hatching and again at repeated 'tidal' intervals thereafter. Thus crab larvae which are an important component of coastal plankton not only behave as tidal clocks, but possess an inherited capability to express such rhythmicity, the timing of which is set by the behaviour of the parent female at the time of hatching. They are hatched as tidal clocks.

The original paradigm that animal plankton species in the sea universally undertake vertical migration rhythms of daily periodicity, now seems to require amendment. Not all species, and not all life-stages of individual species, swim upwards at night, some even undertaking downwards migrations at that time. Moreover, some coastal plankton species, and some estuarine species as will be seen in the next chapter, undertake vertical migration rhythms of tidal, not only daily periodicity. Importantly too, it has come to be recognized that such rhythms may be partly under the control of internal biological clocks and are not driven solely by environmental factors such as light. However, further investigation of plankton animals as living clocks will only be achieved when present difficulties have been overcome in carrying out experimental studies on a greater range of ocean plankton species over long periods of time, either in shore-based laboratories or aboard research ships at sea. Ultimately too, fuller understanding of the geographical dispersal of marine and estuarine animals must also take account of clock-driven vertical migration rhythms, as will also become apparent in the next two chapters.

8

Staying put in estuaries

...the estuarine environment is more extreme, and undergoes more violent fluctuations than the open sea or freshwater habitats.

Eric Perkins, 1974

Throughout the world, where rivers flow into the sea and meet the pulsing of ocean tides, living conditions are hazardous for aquatic organisms. Marine fish such as salmon which enter freshwater to spawn, and freshwater eels that spawn in the sea, face the problems head on as they move through the admixture zone where salt and fresh waters meet. Such anadromous and catadromous migrants have evolved physiological mechanisms by which they are able to cope with dramatic changes in water flows and the rapid uptake or loss of salt and water to which they are exposed, depending upon their direction of travel through the salt and freshwater interface (Davenport, 1985a, b). Fully marine or freshwater animals do not have such compensatory mechanisms so, if they are inadvertently carried into the admixture zone, their survival is threatened. However, some opportunistic species have successfully colonized estuarine habitats, permanently overcoming the problems of fluctuating salinity, temperature and water movements to become estuarine residents. Some macroalgae are specific to estuaries, remaining attached to rocky outcrops, quays and piers, and large populations of microalgae are to be found in the upper layers of mudflats, some migrating deeper at night under the control of biological clockwork, as discussed in Chapter 2. Among animal species, the high energy costs of coping with the problems associated with survival in estuaries are seemingly outweighed by the benefits arising from the fact that relatively few have taken the evolutionary risk of colonizing such hazardous environments. In comparison with the densely populated environments of the sea and freshwaters, living conditions of animals in estuaries offer fewer risks from predators and lower competition for food and space (McLusky, 1981).

Naturally, the problems for estuarine organisms vary according to the volume of freshwater discharged at the river mouth and the extent of the tidal oscillations that repeatedly force seawater upstream against the river outflow. For animals, central to the requirements for residence in, or passage through estuaries has been the evolution of physiological mechanisms of osmoregulation that permit them to cope with sudden fluctuations in the salt content of the surrounding water (Lockwood, 1976; McLusky, 1981; Davenport 1985a). In addition, some have also acquired behavioural mechanisms that permit them to avoid particularly hazardous conditions (Davenport, 1985b). For example, common green shore crabs *Carcinus maenas* often extend their range into British estuaries but, at certain stages of their moulting cycle they are able to avoid unfavourably low salinities by the behavioural response of halokinesis (Thomas *et al.*, 1981; Ameyaw-Akumfi and Naylor, 1987b; McGaw and Naylor, 1992). Crabs that are soon to moult show greatly increased locomotor activity when exposed to dilute seawater, a response which increases their chances of survival by stimulating them to move to less dilute seawater in a salinity gradient. Such behaviour explains how it is that newly moulted, soft-shelled crabs are found less commonly high up in an estuary, where they would be more vulnerable to salt loss and water uptake than is the case when their shells are hardened. Conversely, the related species of crab, *Carcinus mediterraneus*, is found localized near estuary mouths such as those of the delta of the River Ebro in Eastern Spain and is usually absent from open coasts around the Mediterranean Sea where salinities are slightly higher than on Atlantic coasts of Europe. This species shows increased locomotor activity in water of high salt concentrations similar to those of the Mediterranean, a response which probably explains its ability to avoid the open sea and to aggregate in delta mouths and estuaries around the coastline of the Mediterranean Sea (Warman *et al.*, 1991b).

But mobile estuarine animals have more to cope with than fluctuations in the salt content of surrounding water. They are often also exposed to the strong currents generated by tides and river flow. Simply maintaining position in an optimal part of an estuary presents such animals with serious problems. For relatively large walking animals such as bottom-living crabs, the problem can be alleviated by clinging to the bottom in strong currents and by walking long distances to avoid adverse conditions. However, for species in estuaries which swim for all or part of the time, position maintenance presents far greater problems.

Notwithstanding tidal ebb and flow, the net transport of water and of animals that swim passively in that water, is seawards driven by the uni-directional flow of the river water. The question therefore arises as to how resident populations of weak-swimming plankton animals succeed in remaining in an estuary, without being carried out to possible death in the open sea. Equally, and of concern to the management of some commercial fishery activities in estuaries is, for example, the question as to how commercial stocks of bottom-living crabs and oysters are sustained. Each of these two groups of animals reproduces by releasing into the water column planktonic larvae, the swimming powers of which are trivial in the context of the strength of estuarine water currents. The larvae are therefore seriously at risk of being swept out to sea as soon as the adults have spawned. Some answers to these questions have been provided, arising partly from scientific curiosity, but also driven by the need to understand, and hopefully predict, sustainability patterns for commercial fisheries in estuaries. The answers derive from an understanding of the behaviour of the animals concerned in relation to the river flows and tidal regimes in the estuaries that they inhabit.

Estuaries transport to the sea large volumes of freshwater that run off from the land. The runoff may be more or less constant or highly seasonal depending upon the size of the river catchment area and the prevailing climatic conditions. Freshwater is less dense than seawater, upon which it tends to float if the two flow together with no stirring, as occurs in estuaries where the tidal range is small. In such stratified estuaries river water flows seawards on the surface, passing over a wedge of denser seawater that pulses in and out beneath, with little mixing at the interface between the two water masses. In contrast, where tidal oscillations are extensive, seawater beneath the freshwater runoff behaves quite differently; it surges strongly in and out with each tide, causing mixing of sea and freshwater, often partly restricting the seaward flow of freshwater as it does so. Depending upon whether estuaries are stratified or mixed they therefore present a number of additional hazards for plankton animals that live there. Even if animals are adapted physiologically to cope with differences in salinity to which they might be exposed, the complicated water flows may push them high upstream or wash them out to sea where, in either case, their survival would be jeopardized.

*   *   *   *

As early as the beginning of the twentieth century the problem of population stability of estuarine animals was recognized by European fisheries biologists concerned with managing stocks of the flounder *Platichthys flesus*. This is a fish of North East Atlantic coasts, which ranges from the

Black Sea through the Mediterranean and northwards to the Baltic and White Sea. Though defined as a euryhaline species that is physiologically adapted to live in localities of fluctuating salinity, the flounder has not achieved complete independence from the sea. Adults and larvae live in salt water and it is only as juveniles that they are able to enter estuaries. In the North Sea, mature adult males and females are found in open waters where they lay and fertilize their eggs, which eventually hatch as free swimming larvae into the marine plankton. Only after the larvae settle to the bottom as juveniles, in the form of miniature adults about a centimetre in length, do they enter their estuarine nursery areas. There they feed and grow until eventually migrating out to sea again, where they become sexually mature and breed to complete the life cycle. Dutch biologists recognized this life history in the early 1900s, with Ehrenbaum (1911) asking how it was that young flounders made their way from the sea into estuaries. But it was not until much later that another Dutch biologist provided a comprehensive answer. Jager (1998) began by sampling coastal plankton at various stages of the tidal cycle at times of the year when free-swimming flounder larvae were known to be abundant. It was known from earlier studies that flounder larvae settled out as juveniles on tidal flats along the Dutch coastline, inshore from the adult spawning stock. Hence it was possible to make extensive collections of the settlers, from which it soon became apparent that arrival inshore of the settling larvae varied according to the tidal cycle (Jager, 1998, 1999a, b). Passive transport inshore by tidal action was excluded because larvae swept inshore on one rising tide would be expected to be carried out to sea again on the next ebb. An active process seemed to be involved when it was found that flounder larvae varied their depth in the water according to the state of the tide. They clearly migrated actively up to the sea surface and down again at different states of tide. During ebbing tides larvae were found near the sea bed in water that moved only sluggishly offshore because of the frictional drag of bottom sediments. They avoided surface waters which flowed much more rapidly offshore as the tide receded. In contrast, it was observed that the larvae tended to swim up towards the sea surface at the end of the ebb tide and to remain there during the first hours of the flood tide in freely flowing surface water which pushed them inshore and into the mouths of estuaries (Figure 8.1). Remarkably, if the larvae were not quite ready to metamorphose into the juvenile flatfish state and were unable to settle on the bed of the estuary, many swam upwards again on next ebb tide. This behaviour ensured that they were flushed out to sea again and their attempts to settle in the estuary were repeated by swimming upwards again on a later flood tide.

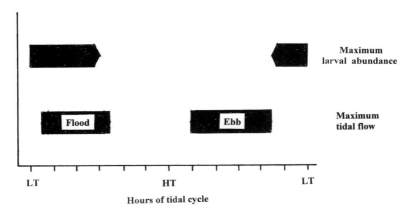

Figure 8.1 Diagrammatic representation of times of greatest abundance of free-swimming flounder larvae in Dutch estuaries compared with the times of maximum flood and ebb of tides (based on data in Jager, 1999b).

The free-swimming larvae appeared to behave as living tidal clocks, intermittently 'floundering about', as Jager (1999a, b) writes, trying not to sink during rising tides. They therefore selectively exploit different tidal flows to achieve their objective of settlement in their estuarine nursery grounds. This process of 'selective tidal-stream transport' is a widely recognized behavioural trait in estuarine and marine invertebrates and fish, including some burrowing species which intermittently emerge and swim in the water column (see Morgan, 1965; Sato and Jumars, 2008). On a large scale, the mysid crustacean *Neomysis americana* in the Damarascotta River estuary in Maine, USA, was observed in autumn to emerge from bottom sediment and swim at tidal (12.4 h) intervals. During low slack tides the mysids moved rapidly into the water column, including near-surface layers where seawards transport was possible. In contrast, slightly later, during peak flood tides they remained near the bottom sediment, where transport into the estuary was more likely. They therefore exhibited a plausible mechanism whereby they were retained in the estuary by selective tidal stream transport (Sato and Jumars, 2008). On a smaller scale, Morgan (1965) observed that individuals of some populations of the amphipod crustacean *Corophium volutator* in a British estuary also emerged at certain times of day from the muddy substrate in which they normally burrow to swim in the water column above. By collecting amphipods with a hand net he showed that quite large numbers could be caught during the flood tide, the numbers reducing significantly at the time of high tide, but rising dramatically again during the ebb tide. Such a symmetrical pattern of swimming in relation to tidal phase appears to be the basis of a clear mechanism involving selective tidal-stream transport that ensures stability of the *Corophium* population at its preferred location within the estuary.

In many cases the rhythmic patterns of swimming shown by estuarine species using selective tidal-stream transport have been shown to be at least partially under biological clock control (Wooldridge and Erasmus, 1980; Gibson, 2004), though whether this is the case with larval flounders and mysid crustaceans referred to earlier remains to be seen. Clear demonstration of endogeneity is only apparent following experiments such as those carried out on *Corophium* by Morgan (1965). When recording the spontaneous swimming behaviour of *Corophium* in constant conditions in the laboratory it was found that the amphipods were most active during expected ebb tide, expressing a temperature-independent, circatidal rhythm which persisted for about three days under constant conditions in the laboratory. This asymmetrical pattern of endogenously controlled swimming clearly differs from the symmetrical pattern of flood- and ebb-swimming observed on the shore, with peak swimming during the rising tide, evidently induced exogenously by tidal immersion. Morgan (1965) further concludes that amphipods carried inland at the edge of the rising tide would be at risk of stranding at high tide when field collections and experimental results suggest that they show a marked reluctance to swim. However, tidal-stream transport downshore is assured by biological clock-controlled swimming during the early ebb tide, the circatidal swimming rhythm probably being reinforced by the decreasing hydrostatic pressure.

* * * *

The problem of population stability of resident adult estuarine animals that release free-swimming larvae was first addressed in relation to commercial exploitation of oysters and crabs in the weakly tidal, stratified estuaries of the eastern seaboard of the United States of America. The planktonic larvae of some species were found to remain in estuaries throughout their period of development until they metamorphosed into the adult stage and joined the parent stock on the bed of the estuary. Surprisingly though, larvae of some other species were found to leave the estuary, become widely distributed in the open sea, yet still find their way back to the estuaries to settle out amongst the resident adult population. Maintenance of the commercial stocks of the organisms concerned therefore depended upon one of two processes that were in need of explanation: larval retention in an estuary and larval recruitment back into an estuary from the open sea.

A long-standing hypothesis concerning the commercially fished populations of oysters in east coast estuaries of the United States of America, initially postulated by Nelson (1912), proposed that the stocks were maintained, despite intensive harvesting, because of the behaviour of their free-swimming, planktonic larvae. It was suggested that the oysters concerned, *Crassostraea virginica*, released larvae which swam upwards into the water column during rising tides, when seawater flooded into the estuaries, but sank and clung to the estuary bed during ebb tides when they risked dispersal to the sea. That hypothesis remained speculative for over half a century until it was tested in the James River estuary, Virginia, by Wood and Hargis (1971). They began by considering the relative speeds of larval swimming and estuarine currents. Lateral swimming by the larvae was quickly excluded as a mechanism for maintaining station, because horizontal tidal currents in the estuary reached 80 cm s^{-1}, which is 80 times faster than the larvae were observed to be able to swim. However, even with a swimming speed of only 1 cm s^{-1} it was apparent that larvae could, hypothetically at least, rise vertically through 9 m of water in 15 minutes. If so, this would be sufficient for the larvae to swim upwards from near the bed of the estuary well into the wedge of sea water that was forced into the estuary below outwards-flowing river water, at the times of rising tide.

To test the hypothesis, it was first required to show that upwards swimming by oyster larvae did occur and that such vertical migrations coincided with the times of rising tides. The opportune time to seek answers to these questions seemed to be when middle stage oyster larvae remained in the estuary after the original mass release of larvae into the plankton above the oyster beds. Many newly released larvae would no doubt have been washed out to sea in estuarine surface water, never to return, but those that remained were sampled at various depths, at regular intervals throughout a day and night. This was carried out from a fleet of five small research vessels stationed in the vicinity of the James River Bridge, somewhat downstream from the main oyster beds. During ebb tides it was found that greatest numbers of larvae tended to remain very close to the bottom, where frictional effects of bottom sediments were such that water flows were minimal. By this means the larvae evidently avoided being carried out to sea during low tides. On the other hand, large concentrations of larvae were found off the bottom in the mass of water forced into the estuary during rising tides (Figure 8.2). That finding raised the question as to whether oyster larvae were carried up into the water column by the stirring action of rising tides, or whether they actively swam upwards at those times. Stirring up of larvae as if

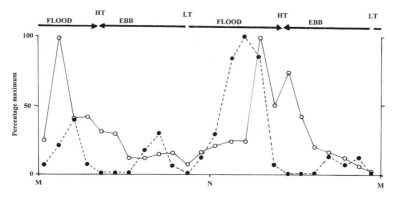

Figure 8.2 Hourly densities of oyster *Crassostraea virginica* larvae (open circles) and coal particles (closed circles) rising above the seabed of the James River estuary, USA, during two tidal cycles. Coal particles were swirled up during flood and ebb tides, but oyster larvae swam up only during flood tides (redrawn from Wood and Hargis, 1971).

they were passive particles seemed unlikely since they were not swept upwards during ebb tides when turbulence would have been as great as during flood tides. But was it possible to confirm that this interpretation of events was correct?

It so happened that coal had been transported by shipping along the estuary in earlier times, resulting in large concentrations of small coal particles in the sediment. Those inert particles were of about the same size as oyster larvae and they were repeatedly stirred up into the water column by tidal currents. They could easily be recognized as such and it was possible to count them along with the oyster larvae collected in estuary water pumped from various depths and filtered aboard the research vessels. Unlike oyster larvae, coal particles were found in the plankton pump samples taken during both flood and ebb tides (Figure 8.2). Their densities in suspension varied, presumably due to differences in the prevailing hydrographical stirring conditions, but there were striking differences between the pattern of upwards swirling of coal particles and oyster larvae. Also, the peaks of abundance in the water column of oyster larvae did not coincide with those of the inert coal particles. The findings suggested that the larvae were behaving as active, not passive particles, choosing, as it were, to swim up only during flood tides. By remaining on the bottom during ebbing tides and swimming upwards into the flooding tides, the net transport of larvae would be upstream. Having initially been released into seawards-flowing surface water which would tend to carry newly released larvae downstream, their quickly acquired tidally related pattern of sinking and upwards swimming behaviour would lessen the risk of being washed out to sea. Behaviour of the larvae, adapted to the complex water circulation within their environment, thus ensures that oyster stocks in the estuary are normally continuously replenished by newly settling larvae. So, oyster populations are evidently maintained by the process of larval retention which is effected by larvae that behave as living tidal clocks. So far, the simplest explanation for their behaviour is that they migrate upwards in the water column in response to the influx of saline water during the rising tide. It remains to investigate the possibility that oyster larvae possess circatidal clocks by which spontaneous upwards swimming would be timed even in the absence of an influx of saline water.

In addition to oysters there are also large populations of commercially fished blue crabs *Callinectes sapidus* in estuaries of the east coast of the USA. The stocks of these crabs are, like oysters, maintained naturally, but by quite different biological processes. Whilst adult blue crabs are able to reside successfully in those estuaries their young larvae are unable to survive in the fluctuations of sea and fresh waters that their parents can tolerate. They require to be exposed to the continuously saline conditions of the open sea in order to undergo their early development. Accordingly, egg-bearing females that are about to shed their young larvae migrate to the estuary mouths, where larval release takes place. Once released from their parents the larvae are rapidly dispersed seawards into optimum conditions required for their early development (Epifanio *et al.*, 1984). Laboratory studies (Sulkin *et al.*, 1980; Sulkin, 1984) have shown that offshore dispersal of first stage larvae of blue crabs is achieved by complex behavioural responses to external conditions. They were shown to swim upwards against the force of gravity, in response to increased pressure, reduced temperature and increased saltiness, all of which they would experience at depth. Upward swimming behaviour induced by these environmental signals would take the larvae to surface waters which, driven by the outflow of river water, would transport them seawards. After seven moults and an increase in size after feeding in the sea, the larvae then showed a greater tendency to sink and their behaviour was found to be quite different from when they were newly hatched. At that stage when tested in the laboratory, they were shown to respond positively to gravity, a response that induced them to sink which, in the sea, would ensure that they found themselves near the seabed in a water mass that would move them inshore again at the start of a journey back into an estuary. Along the Atlantic coasts of the USA bottom water tends to drift landwards, forced by tides and winds, particularly during the hurricane season (Scheltema, 1975), resulting in the arrival of large numbers of blue crab larvae into the mouths of estuaries. There the sea-going zoea larvae moult into the megalopa stage, which make their way into estuaries before metamorphosing finally into juvenile crabs in the vicinity of the parent stocks. In seeking to forecast the annual recruitment strength of megalopa larvae to the commercial stocks of blue crabs in US east coast estuaries the ingress has been modelled and related to the pattern of original larval release from the parent stock (Tilberg *et al.*, 2008).

The successful recolonization of estuaries by the megalopa larvae of *Callinectes sapidus* is thought to be achieved by further responses to water pressure changes experienced at different depths. Naylor and Isaac (1973) showed that blue crab megalopa larvae showed little response to increases in hydrostatic pressures which they would experience down to depths of 4 m or so, but showed markedly increased swimming activity in response to pressures equivalent to depths below 4 m. By this means it was inferred that the megalopae would sink below the seawards-flowing water at the surface of a stratified estuary, but would swim in the tidally pulsed bottom water which would carry them upstream. Further development of this conceptual model for flood-transport into an estuary suggests that megalopa larvae ascend into the water column in response to an increase in salinity (Tankersley *et al.*, 1995) and remain swimming in response to water turbulence (Welch *et al.*, 1999; Welch and Forward, 2001). This behaviour would again enhance the chances of larvae being carried into estuaries during rising tides. Finally, it has been shown that when the megalopae are ready to metamorphose and sink on to the bottom as juvenile crabs, they do so in response to decreased turbulence during slack water at the end of the flood tide that carried them to their final destination (Tankersley *et al.*, 2002).

So, just as oyster larvae show tidal rhythms of behaviour when maintaining position in an estuary, blue crab larvae also do so when entering an estuary after dispersal seawards. Beyond this intriguing behaviour, however, blue crab larvae also exhibit a fortnightly, that is semilunar, pattern of behaviour. This was demonstrated by placing artificial collectors in the Newport River estuary, North Carolina, on which it was hoped that blue crab larvae would settle at the end of their journey upstream. Hopes were rewarded and it was discovered that larvae settled in greatest numbers during neap tides at the lunar quarters (Forward *et al.*, 2004). Semilunar periodicity of settlement phased to neap tides in this way is thought to have evolved because tidal transport of larvae is most effective when rising tides occur during the first 6 h or so of darkness at night.

In the Newport River area this particular timing of tides occurs during neap tides, when the end of the flood tide period occurs after midnight.

*   *   *   *

It remains to be established whether or not oyster larvae and blue crab larvae possess biological timing mechanisms of tidal or semilunar periodicity that enable them to behave as biological clocks. However, tidal clockwork has been clearly demonstrated in the larvae of the diminutive, and therefore commercially unimportant, mud crab *Rhithropanopeus harrisii*, another resident of east coast estuaries of the USA (Forward, 2009). Larvae of this crab collected in the Newport River estuary were studied in laboratory experiments carried out by Cronin and Forward (1979), Forward and Cronin (1980) and Cronin (1982) having first established that free-swimming larvae of *Rhithropanopeus* undertake vertical migrations of tidal periodicity and are retained in the estuary near the resident adult crabs during their entire period of development to the megalopa and adult stages. Larvae were placed in a vertical transparent cylinder containing estuary water maintained for several days at constant temperature and in continuous darkness. They were kept in good condition by feeding them brine shrimp larvae (*Artemia*) at irregular intervals, to avoid conveying to them any time cues related to tidal rise and fall or day and night. Observations of larvae swimming in the transparent cylinder were made by remotely switching on a dim red backlight for two minutes every half hour and scanning them with a television camera which recorded to a tape deck. Subsequent analysis of the videotape showed that the crab larvae survived in substantial numbers for four to five days in the recording chamber, during which time they repeatedly swam upwards *en masse* at the times of high tide that they would have experienced had they remained in their natural environment. They sank downwards again when they would expect to experience low tides, indicative of oscillatory vertical swimming that clearly seemed to be under the control of their own biological clocks (Figure 8.3). In nature the clocks appear to be continuously reset by environmental cycles of light, salinity and current speed but, in the absence of the environmental variables, the clocks are expressed as typical circatidal rhythms of behaviour in the laboratory. In the estuary, the tidal rhythm of vertical migration ensures that the larvae spend half of their time near the surface where the net flow of water is seawards and the other half below mid-depth where, because of tidal action, the net flow of water is into the estuary. Similar rhythms of behaviour are shown by all planktonic stages of development of *Rhithropanopeus*, reducing the risk of their transport out of the estuary and ensuring their recruitment back to the resident adult crab population. Later, it was shown that the larvae of two species of fiddler crabs *Uca*

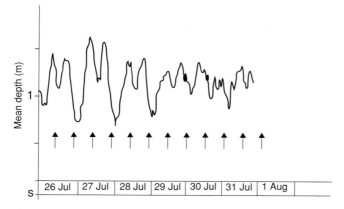

Figure 8.3 Three-hourly positions of the centre of distribution of a group of freshly collected *Rhithropanopeus* larvae maintained in a vertical column in constant conditions in the laboratory. S, surface; arrows, times of predicted low water at the sampling site (after Cronin and Forward, 1979; by permission of the American Association for the Advancement of Science and the authors).

(Plate 16) also exhibited circatidal rhythms of upward swimming, again as mechanisms enabling them to remain in the vicinity of the parent populations in North American estuaries (see Forward and Tankersley, 2001). These discoveries demonstrated unequivocally the occurrence of circatidal rhythms of vertical migration in estuarine planktonic larvae, suggesting that such animals possess internal circatidal clockwork which enables them to exploit the layered water circulation patterns of stratified estuaries in order to maintain position or recruit back into such estuaries after dispersal seawards.

* * * *

However, it still has to be explained as to how it is that small plankton species maintain their position in estuaries that are strongly tidal, in which sea and freshwater tend to be more thoroughly mixed together and move as one mass. Under these conditions the main differences of flow patterns that can be discerned by the use of current meters indicate that water flows are sluggish near the bottom, whatever the direction of flow, due to the frictional drag of water along the estuary bed, whereas higher in the water column the flows are stronger. The question arises as to whether plankton species are able to exploit these differences of flow strength in well mixed estuaries which lack stratification. Such estuaries clearly present different selection pressures against which different position maintenance or re-recruitment behaviours might be expected to have evolved.

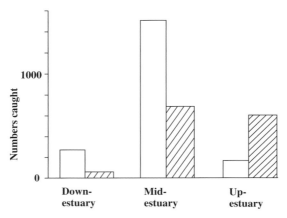

Figure 8.4 Numbers of the copepod *Eurytemora affinis* taken at mid-depth in standard plankton samples during flood (unshaded) and ebb (shaded) tides along the estuary of the River Conwy, N. Wales (after Hough and Naylor, 1991; by permission of Inter-Research).

The problems are serious for plankton species such as the copepod crustacean *Eurytemora affinis*, an abundant crustacean near the base of plankton food chains in many European estuaries. It has been suggested that they maintain position in the water column by actively swimming against the prevailing water currents (Heckman, 1986). Furthermore it has been reported that when they find themselves in their preferred salinity zone in an estuary, they somehow manage to improve their chances of staying there by avoiding adverse currents that would sweep them into waters that were too saline or too fresh (De Pauw, 1973). Unfortunately, such mechanisms seem to invoke swimming capabilities that are much greater than would be expected, or indeed have been recorded, in such small animals (Castel and Viega, 1990). To try to resolve the apparent dilemma of how population stability was achieved, *Eurytemora affinis* was studied in the strongly tidal, mixed estuary of the River Conwy on the coastline of North Wales (Hough and Naylor, 1991). In samples taken by tow nets, the copepod was found to vary in abundance in mid-water at different points along the estuary, according to the state of the tide throughout the day. At the seaward end of the estuary the copepods were found to accumulate most abundantly in mid-water just before high tide, permitting them to exploit flood tides which would carry them upstream. In contrast, at the landward end of the estuary they were maximally abundant in mid-water catches just after high tides, so placing themselves at the mercy of ebb tides which would carry them downstream again (Figure 8.4). These findings suggested that there were consistent phase differences in the times when animals swam upwards from sluggish water near the estuary bed into faster moving water nearer the surface, perhaps dependent upon the salinity of the surrounding water. By this means, copepods found towards the mouth of the estuary would be transported upstream by swimming upwards during rising tides whereas, if they were carried too far into freshened waters, upwards swimming during falling tides would carry them downstream again. Consequently the bulk of the population would tend to accumulate at a mid-estuary position.

Intriguingly, it was also found that catches of *Eurytemora* also varied over the neaps/springs cycle of tides; their zone of greatest abundance occurred higher up the estuary during spring tides than during neaps (Hough and Naylor, 1991). Seawater is clearly pulsed farther into tidal estuaries during spring tides than during neaps and the bulk of the copepod population appeared to respond accordingly, maintaining itself within preferred salinity limits. The neaps/springs changes in position were effected because greatest numbers swam up into the water column on the flood during spring tides, whereas during neaps greatest numbers swam during ebb tides.

The question then arose as to whether the rhythms of upwards swimming by *Eurytemora* occurred only in response to the salt concentration of their surrounding water, or whether again it was timed by internal circatidal clockwork. To address this question the swimming behaviour was observed in copepods maintained in the laboratory in transparent vertical cylinders containing estuary water. Those experiments showed that the rhythms of vertical swimming, which in the estuary affect the horizontal distribution of the copepods, were truly endogenous; they persisted at circatidal periodicity in constant conditions in the laboratory (Hough and Naylor, 1992), exactly as was the case in the newly hatched larvae of shore crabs (Figure 7.4). Moreover, the phase differences in the times of maximum swimming shown by copepods collected at different points along the estuary and throughout the neaps/springs cycle were also replicated in constant conditions in the laboratory (Figure 8.5). *Eurytemora* collected from mid-estuary during spring tides of increasing amplitude showed peak swimming before expected high tide. In contrast, copepods collected at the same time, but near the upper limit of tidal influence, showed peak swimming after the times of expected high tide, as also did copepods collected from mid-estuary when spring tides were decreasing in amplitude. Thus *Eurytemora affinis* not only behave as living tidal clocks, but also possess internal biological clockwork of circatidal periodicity that drives their tidal rhythms of upwards

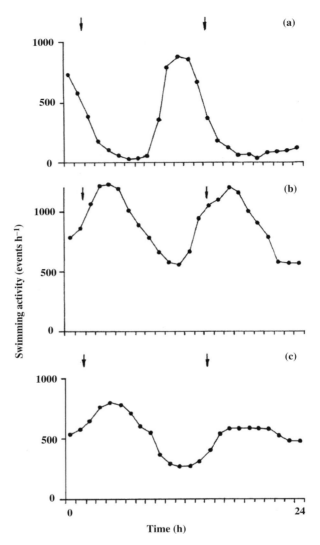

Figure 8.5 Spontaneous swimming activity of *Eurytemora affinis* recorded by infra-red beam interruptions in actograph chambers maintained for 24 h in constant darkness and temperature. Approximately 2000 copepods were used in each experiment, freshly collected from a tidally mixed estuary as follows: (a) from mid-estuary during spring tides of increasing range; (b) from mid-estuary during spring tides of decreasing range; and (c) from near the upper limit of tidal influence during spring tides of decreasing range. Arrows, times of expected times of high tide (after Naylor, 2006; data in Hough and Naylor, 1992; by permission of Taylor & Francis Ltd.).

swimming. The clockwork has probably been selected for during evolution by a requirement for the copepods to survive most efficiently in water that is neither too salty nor too fresh, and it is probably synchronized by environmental cycles of salinity variation. *Eurytemora*, therefore, is another example showing that the behavioural basis of orientation and position maintenance behaviour in an estuarine plankton species is determined by endogenous circatidal rhythms of upwards swimming. However, the mechanism adopted by *Eurytemora affinis* is unlike that of the zoea larvae of the crab *Rhithropanopeus harrisii* discussed earlier, which exploits layered circulations in weakly tidal, stratified estuaries. By contrast, *Eurytemora* maintains position in a tidally mixed European estuary by continuously adjusting the phase of its circatidal swimming rhythm.

From the examples discussed in this chapter it can safely be concluded that several estuarine zooplankton species not only behave as living tidal clocks but do so under the control of their own internal biological clockwork. Their endogenous circatidal clockwork, cued by the moon's influence on tides, appears to have evolved as a means of helping to ensure the survival of such animals in the hazardous conditions that prevail in the estuarine environment.

9

Ocean drifters

Animal plankton [comprises] small, free-swimming animals whose powers of swimming are not great enough to overcome the transporting effects of currents and tides.

John S. Colman, 1950

Assemblages of organisms in the sea are commonly and very broadly classified according to their way of life. Sea floor burrowers, crawlers and fixed forms comprise the benthos, strong swimmers such as fish, mammals and squid are designated as nekton, and the poor swimmers as plankton. The last group, named from the Greek verb to wander, roam or drift, is a particularly heterogeneous assemblage of living organisms ranging from prokaryotic cyanobacteria (so-called blue-green algae) and true unicellular microalgae to large coelenterates, often including the eggs and larvae of benthic invertebrates and of fish. The swimming powers of animals within this motley group are weak, testified by the mass stranding of coelenterates and other plankton species that are often found along coastlines after a spell of onshore winds at certain times of year. Consistent with the view that the swimming powers of planktonic animals are not great enough to overcome ocean currents (Colman, 1950), many early marine biological studies that focussed on the distribution and spread of benthic species, did so on the assumption that their planktonic larvae dispersed passively in ocean surface currents (see Thorson, 1964; Scheltema, 1971). This was not an unreasonable assumption since dispersal benefits encroachment of new localities by distributive larvae, as has recently been quantified by studies of the free-swimming planktonic larvae of shore-living fishes in the Mediterranean Sea. Species with larvae that have a short life span in open sea plankton were found to have a more restricted geographical distribution than species whose planktonic larval life is more protracted (Macpherson and Raventos, 2006).

Indeed, physical oceanographers interested in patterns of ocean circulation have sometimes regarded small planktonic animals as inert particles to be used as markers for the validation of mathematical models of ocean current circulation systems (Naylor, 2006). For example, many plankton species remain together in huge aggregations, giving rise to plankton patchiness at scales of 10–100 km (Steele, 1976). Originally it was considered that such patchiness occurred solely in response to hydrographical forces in the form of swirling ocean currents, but it is increasingly apparent that many such animal aggregations are also formed and maintained because of behavioural interactions between individuals (Ritz, 1994). For too long the term plankton has 'implied passivity, limited swimming power and limited possibilities for sophisticated behaviour because of small size' (Hamner, 1988; quoted by Ritz, 1994). However, behaviour-based aggregations have been demonstrated in copepod, euphausiid and mysid crustaceans (Ritz, 1994), suggesting that the poor swimming ability of many plankton species does not necessarily condemn them to drifting randomly as inanimate objects, completely at the mercy of ocean currents.

But it is already apparent from Chapters 7 and 8 that planktonic animals have some control over their own destiny from earlier discussions of the phenomenon of vertical migration and the ability of estuarine species to stay put without being washed out to sea. In those discussions it was noted that many zooplankton species are able to undertake vertical migrations of daily and/or tidal periodicity, quite independent of water movements around them. These migrations, though at slower speeds than some ocean currents, may be over considerable vertical distances. Just as in estuaries, therefore, oceanic plankton animals may sink or swim from a layer of water moving in one direction to a higher or lower layer moving in another direction, within daily or tidal timescales. Larvae released into the open sea and drifting aimlessly might, fortuitously, settle in a new and suitable locality to permit spread of the species, but many more would die by being eaten or by settling out in unfavourable locations. In estuaries, vertical migrations by planktonic larvae of bottom-living species permit them to exploit layered patterns of water circulation, ensuring that their chances are increased of settling in a favourable locality, thus improving the chances of survival of their species. And where better would conditions be satisfactory for settlement and survival than near the parent stock in the case of crabs and oysters or, for flounders, in larval nursery grounds established over many years of evolution? There is then a paradox concerning the evolution of distributive larvae in coastal and estuarine animals: there is survival value for a species if it is dispersed over a wide geographical range, but there is also selective advantage in settling out from the plankton in optimal conditions for adulthood. The question is whether the same paradox is apparent in the evolution of open ocean species, many of which release planktonic larvae which may be dispersed over vast distances in relation to the sizes of the animals concerned. In such cases, if maintenance of the species necessitated the return of plankton species to areas where they themselves were spawned, then they would be required to undertake very precisely orientated reciprocal journeys, perhaps over considerable distances. The paradox was considered by Longhurst (1976) concerning a population of the planktonic crab *Pleuroncodes planipes*, a population of which appeared to be maintained naturally off the west coast of the southern USA, yet its larvae were known to be transported by the California Current into the open Pacific Ocean, over a distance several thousand miles south west of the normal range of the adult population. It was proposed that the larvae made their way back by sinking into an opposing undercurrent, a hypothesis that was tested by depth-discrete sampling of *Pleuroncodes* larvae. Using a Longhurst–Hardy Plankton Recorder it was 'revealed that vertical, seasonal distribution of larvae and post-larvae was in accordance with that demanded by the hypothesis' (Longhurst, 1976). The larvae did indeed exploit a deep counter current to return to the vicinity of the parental location.

Similarly, it has long been known that plankton animals in the Strait of Gibraltar maintain their horizontal position, despite strong currents through the Strait that would be expected to displace larvae into the Mediterranean Sea or out into the open Atlantic Ocean. Population stability has been demonstrated by echo-soundings reflected by animals of the deep-scattering layer, which were found to show vertical movements in accordance with changing current flows at different depths within the Straits at different times of day (Frasseto *et al.*, 1962). Also, depth regulatory behaviour has been invoked to explain dispersal and recruitment of the deep sea crab *Geryon quinquedens* found off the east coast of North America (Kelly *et al.*, 1982). Newly hatched larvae released by female *Geryon* living on the slope of the continental shelf off Cape Hatteras, in North Carolina, USA, are transported in a south-westerly direction by sub-surface currents flowing in that direction along the continental shelf and down the continental shelf slope. Later in their development larvae are to be found nearer the sea surface and are transported back to the Cape Hatteras region by the Gulf Stream, which flows in a north-easterly direction in that locality (Kelly *et al.*, 1982).

How the detailed behaviour patterns of planktonic larvae of deep water organisms contribute to their patterns of homing remains largely to be investigated. Curiosity-driven research of this kind concerning deep sea animals has been slow to develop, not least because it requires the use of ocean-going research vessels which are expensive to operate. Fortunately though, for marine scientists interested in the seemingly esoteric phenomena of plankton migrations in the ocean, there are examples of larval homing behaviour in some open sea animals that are of commercial value. Funding for behavioural research on those species comes much more readily, on the grounds that it provides background information for the formulation of improved fisheries management schemes for the species concerned. Along the way, such applied research also provides fundamental background information for wider understanding of the coherence of plankton populations in general. For example, a study in the western Irish Sea (White *et al.*, 1988) sought to understand possible controls on the level of recruitment to commercially fished populations of the Norway lobster *Nephrops norvegicus*. Recognizing that the distribution of the adult lobster population is tightly controlled by their requirement for a muddy substrate in which to construct their burrows (Farmer, 1974; Atkinson and Naylor, 1976), the pattern of release of their planktonic larvae was investigated in relation to concurrent oceanographic circulation. White *et al.* (1988) demonstrated that distributions of *Nephrops* larval stages I–III showed a pronounced tongue of high numbers extending southwards from the deep muddy locality where they were hatched into density-driven water flows at the time of release. Even many stage I larvae were found well to the south of the adult population, raising the question as to the fate of the larval population as a whole during the advection process. Many no doubt fall prey to other organisms and are completely lost to the system, evidenced by the lack of *Nephrops* populations to the south. Little is known of the behaviour of stage IV (post-larvae), which settle out as juveniles and possibly migrate along the bottom back to the adult locality, as is known to occur in juvenile edible crabs *Cancer pagurus* over distances of 50 km in the North Sea (Nichols *et al.*, 1982). However, the most likely explanation for the survival of the adult population is that it depends upon sufficient larval retention despite the massive advective losses from the system (White *et al.*, 1988). A fuller understanding of the retention process, as a basis for *Nephrops* fishery management considerations in the western Irish Sea, is clearly dependent upon better knowledge of the variability of coastal currents in that locality and the behaviour of Norway lobster larvae therein.

* * * *

Extending over about 600 km of the coastline of Western Australia, north of the city of Perth, are resident populations of adult western rock lobster *Panulirus cygnus*, which are the resource of a highly valuable commercial fishery. Up to 200 or more kilometres offshore from the coastline, in the south east Indian Ocean, are to be found vast numbers of the planktonic larvae of the lobsters. How do the larvae get there and, more importantly for lobster fishermen, how and in what numbers do larvae return to settle and replenish the stocks? These are fundamental questions for Australian fisheries scientists, who are required to provide advice for sustainable management of the lobster fisheries. Are the larvae merely drifting passively and thereby randomly dispersed by ocean currents, with only a few of the large numbers of larvae released at hatching eventually settling by chance near the parent stock? Or do they exhibit behaviour as a result of which more larvae return than would be expected to do so by chance?

Intensive study of the developmental biology of *Panulirus cygnus* has shown that hatching of larvae from individual egg-bearing females takes place at night over a period of three to five days in the Australian summer months of November to February (Phillips, 1981). The rock lobsters, sometimes called spiny lobsters, crawfish or marine crayfish, have particularly long-lived phyllosoma larvae, so-called because of their leaf-like body shape. They survive in the plankton for eight to eleven months, journeying westwards into the Indian Ocean, moulting eight times and growing as they do so. By the time they reach the ninth phyllosoma stage, ready to moult to the puerulus stage and then to the first tiny lobster stage, many more than would be expected by chance have returned to the rocky coastline inhabited by the parent stock.

Soon after hatching the early phyllosoma stages can be caught in large numbers near the surface of the sea at night, regardless of moonlight intensity. However, by day they are absent from the surface layers and can only be collected at depths of 30–60 m (Rimmer and Phillips, 1979). They clearly undergo daily vertical migrations that are typical of many oceanic plankton species (Figure 9.1). At night they occur at the sea surface in water that is driven westwards by offshore winds. The winds abate or reverse and blow onshore by day but, by that time the phyllosomas migrate downwards to depths unaffected by surface winds. The net transport of the larvae in their early stages is therefore westwards into the south east Indian Ocean where they feed and grow. The larvae clearly

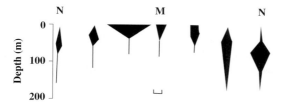

Figure 9.1 Density in relation to depth of early stage phyllosoma larvae of rock lobsters *Panulirus cygnus* off Western Australia. Upward migration in the ocean occurred at night. Scale bar indicates 100 000 larvae per cubic metre of sea. N, noon; M, midnight (after Rimmer and Phillips, 1979; by permission of Springer Scientific and Business Media).

behave as living clocks, exhibiting a daily rhythm of vertical migration by which they achieve dispersal westwards. Whether they possess internal biological clocks or migrate vertically solely in response to the day/night cycle has yet to be ascertained.

By the time the phyllosomas reach the later stages of their development, the depths to which they descend by day are much greater than the depths achieved by the early larvae; they are often found at depths of 120 m. Moreover, though they continue to show daily vertical migration behaviour, they concentrate at the sea surface only on nights when moonlight intensity is less than five per cent of that of full moonlight. Therefore on many nights, and during all days, they are concentrated well below the surface layers of the sea. In this situation, at their late stage of development, the larvae largely avoid wind-driven offshore currents and spend more time in a deeper, coastward-flowing, current that brings them inshore towards the coastline where they were originally spawned (Phillips, 1981).

There is little doubt that with a long development time and dispersal over a wide geographical area there is extensive mortality of phyllosomas, by predation or by the vagaries of weather and ocean currents. Nevertheless the larvae are not behaving as passive particles that are dispersed randomly in the ocean. Recruitment back to the parent stocks, by more larvae than would be expected by chance, clearly maintains the commercially fished stocks of lobsters along the western Australian coastline. It is certainly understandable how such a lifestyle could have evolved gradually over geological timescales. Given the occurrence of lobsters that release vertically migrating planktonic larvae into an ocean locality with relatively consistent current patterns, natural selection would favour larvae that had an extended development time and were able to exploit ocean currents. They would be able to utilize their clock-driven patterns of behaviour in order to disperse seawards for feeding purposes, yet to return to the locality of the parental stock at the time of their crucial moult to the miniature adult stage.

A further example of the seemingly non-random wanderings of marine plankton relates to the larvae of a population of the edible crab *Cancer pagurus* off the coasts of southern England at the eastern end of the English Channel. In spring, berried females of this crab, carrying mature eggs, release zoea larvae which immediately swim towards the sea surface and are transported by the residual surface flow of water eastwards towards the Dover Strait. The larvae eventually sink to the bottom again and metamorphose into bottom-living crabs that at first might be thought to be lost from the parent population to the west. However, this is not so. Fishery scientists collected crabs in the English Channel, attached tags to their shells, released them back into the sea at the point of collection and awaited notification of their eventual recapture by commercial fishermen (Bennett and Brown, 1983). In virtually all cases recaptured marked crabs, particularly adult females, were found to have made extensive migrations of up to 200 km westwards from their release points until their time of capture by fishermen. Adult crabs clearly migrated back towards the original breeding locality where they were spawned from their parents as larvae. Migration by the adults achieves completion of the life cycle of the edible crab and compensates for the dispersal track of newly released larvae.

The question then arises as to how unidirectional transport of the larvae of *Cancer pagurus* in the English Channel, as distinct from random dispersal, is achieved. Though prevailing winds are from the west in that locality, it appears that the crab larvae, unlike rock lobster larvae off western Australia, are not exploiting wind-driven currents at the sea surface to travel eastwards. There are particularly strong tides in the area and it seems that the larvae are using tidal currents accordingly. In some way their daily patterns of vertical migration, which are probably biological-clock controlled, seem to permit them to utilize tidal currents in the eastern English Channel to take them in an easterly direction. It is at first puzzling to consider how vertical migrations of daily periodicity might combine with current flow changes of tidal periodicity to effect larval transport in one direction, but computer simulations of what might happen in such circumstances have shed light on the problem.

* * * *

As discussed in an earlier chapter, vertical migration has long been recognized as a potentially important mechanism by which planktonic animals in estuaries can regulate their horizontal position. Equally, as also

seen earlier, if the period of a vertical migration rhythm is exactly synchronized with the tidal period, a particularly efficient transport system
ensues. This mechanism of selective tidal stream transport, whereby animals swim up to the surface during successive flood or ebb tides and
swim downwards at the opposite phase of tide into more static water
towards the sea bed, ensures unidirectional transport during reversals
of tidal currents. Recruitment and retention of plankton in estuaries
(Chapter 8) and the dispersal offshore of the larvae of the common shore
crab *Carcinus maenas* (Chapter 7) exemplify this phenomenon.

There is, however, a problem in considering horizontal transport
of the larvae of organisms such as the edible crab *Cancer pagurus* which
undertake vertical migration rhythms of daily periodicity in the open
waters of the continental shelf. A larva which undertook vertical migrations of daily periodicity in currents that are driven by moon-generated
tides would simply oscillate about its release point, due to tidal ebb and
flow, and would experience no net horizontal displacement over a semilunar (14.8-day) cycle. However, ocean tides are not generated by lunar
gravity alone; the sun also generates a gravitational, albeit smaller, pull
recognizable as a solar component of tides which has a period of exactly
12 h, determined by the rotation of the earth. This tide interacts with
the moon-generated tide, sometimes pulling against it, at neaps, and
sometimes exaggerating it, at springs. So, plankton species which undertake vertical migration over the 24-h day, though beating out of phase
with lunar tides, would migrate up and down precisely in phase with
the solar component of tides. For many years, little consideration was
given to the fact that vertical migration of daily periodicity by a plankton
animal would exactly match the solar component of tides. To make this
point computer models have been generated of larval behaviour in a tide
driven solely by lunar gravity, in a tide driven solely by solar gravity and
in a mixture of the two (Hill, 1991a, b, 1994). Because the gravitational
pull of the sun is slightly less than half (46%) of that of the moon, tidal
currents caused by the sun are less than those caused by the gravitational pull of the moon. Taking into account these differences in solar
and lunar generated tidal currents, it was then possible to plot the horizontal displacement of hypothetical larvae, depending upon whether or
not they were assumed to have undertaken a daily rhythm of vertical
migration. The model was then run to cover successive 30-day periods of
predicted larval movements in a variety of hypothetical tidal conditions,
each covering a full lunar cycle.

In the model simulations (Figure 9.2) non-vertically migrating larvae oscillated horizontally for 5 km or so about their release point for the

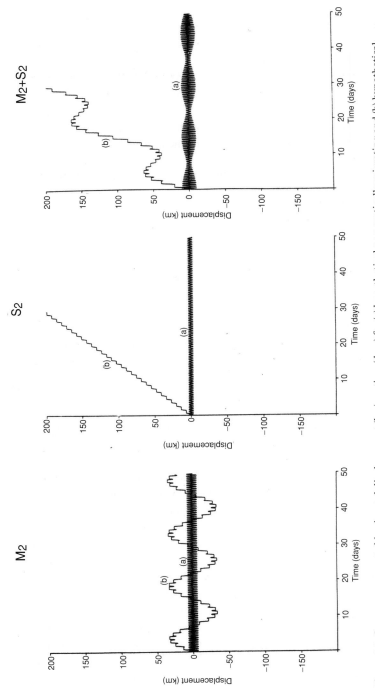

Figure 9.2 Computer generated horizontal displacement (km) vs. time (days) for (a) hypothetical non-vertically migrating and (b) hypothetical 24 h vertically migrating plankton organisms in M2, S2 and M2+S2 generated tidal currents. Assumptions are: M2 currents, 1 m s^{-1}; S2 currents, 0.5 m s^{-1}; organisms spend 6 h in the flow and 18 h on the sea bed where flow is zero (after Hill, 1994).

whole 30 days, whether in moon-generated tides (M2), sun-generated tides (S2) or in both combined (M2+S2). However, the results were quite different in computer simulations in which larvae were assumed to show a 24-h rhythm of vertical migration, spending 16 h near the sea bed where currents are negligible, but rising up for 8 h at the same time each day into tidally induced currents near the surface. Near the surface, larvae were affected differently by tidal currents depending upon whether the computer model simulated moon-generated, sun-generated or mixed moon and sun-generated tides. In moon-generated tides, vertically migrating larvae again oscillated horizontally about their release point, but in those cases for as much as 30 km each way. In contrast, daily vertically migrating larvae subjected only to a solar tide drifted with a residual current in one direction for 200 km. When the 'real' situation was imposed by exposing hypothetical vertically migrating larvae to combined lunar and solar tides the linear displacement was again about 200 km, though with considerable oscillations backwards and forwards along the way. This kind of tidal regime in the eastern English Channel could clearly explain the eastwards drift of *Cancer pagurus* larvae towards the Dover Strait, provided the larvae behaved as living clocks, as they seem to do, showing daily vertical migration between surface waters affected by tides and deeper layers that are slowed down by bottom friction.

Tidal currents at any locality vary with the state of tide and time of day, but unlike wind-driven currents, they nevertheless represent a highly predictable component of flow in the ocean. From an evolutionary point of view this predictability reflects a highly stable part of the physical environment, which may permit natural selection of behavioural strategies that provide adaptive advantage to animals that live there (Hill, 1994). There are geographical differences in the phase, strength and direction of tidal currents that are generated by the sun's gravity, which result in related variations in the horizontal dispersal of vertically migrating plankton organisms dependent upon geographical locality. Moreover, from further computer simulations (Hill, 1994), the direction of net transport of larvae was shown to depend upon the phase of the solar-generated tidal current cycle in relation to the timing of local noon. The phase is fixed for any particular geographical location and seems to be responsible for creating regions in the open sea that are favourable for either retention or directional dispersal of daily migrating animal plankton (Figure 9.3). The differences perhaps determine why some localities in the sea are favoured as spawning grounds and others are not. One such region of directed dispersal occurs in the eastern part of the English

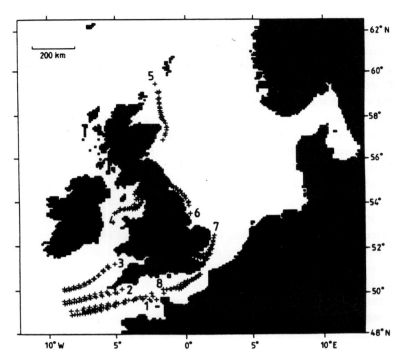

Figure 9.3 Computer-generated 100-day dispersal tracks of hypothetical plankton animals exhibiting 24 h vertical migration rhythms in simulated natural conditions of moon- and sun-generated tides around the British Isles. Dispersal tracks start at the position of each numeral. Note the regions of divergence and convergence (after Hill, 1994; by permission of Elsevier).

Channel where it has been shown that eastwards drift of larvae of the edible crab *Cancer pagurus* appears to compensate for westwards migration by the adult crabs. Seemingly, natural selection has favoured the gradual and concurrent evolution of larval behaviour that encourages dispersal in one direction and adult migratory behaviour in the reverse direction, at all times ensuring completion of the life cycle.

Computer simulations have also been used to suggest that larvae of the crab *Carcinus maenas* returning from waters of the open Atlantic Ocean to Portuguese estuaries probably do so by the use of diurnal vertical migration behaviour (Marta-Almeida *et al.*, 2006). In simulations of this behaviour in relation to ocean currents it was shown that 'larvae' without daily migration behaviour would be advected offshore. In contrast, those undergoing 24-h vertical migration rhythms would be expected to spend a significant amount of time in inshore-flowing bottom water forced by upwelling of deep water at the edge of the continental shelf. Moreover, in a test of such flood-tide transport hypotheses, it has been shown that planktonic larvae of the common mussel *Mytilus edulis* exploited tidal currents as they returned inshore to settle after dispersal in coastal waters (Knights *et al.*, 2006). Larvae were found to be homogeneously distributed throughout the water column during flood tides but to occur in dense aggregations in middle and bottom waters during ebb tides. The net direction of transport was inshore in the direction of flood tides. Among other species that have been studied in this regard is the commercially important plaice *Pleuronectes platessa*, which is fished extensively in the North Sea. Larval stages of this flatfish that are released into the plankton tend to drift away from the spawning grounds, to which as adults they must eventually return to spawn. The problem of how return to the breeding grounds is achieved has been solved by capturing plaice, fitting them with tiny depth recorders, returning them to the sea and following their movements using tracking devices aboard a research ship (Metcalfe *et al.*, 1990, 2006; Hunter *et al.*, 2004). It was shown that when the fish return to the spawning grounds they swim up from the bottom during flood tides and remain on the bottom when the tide ebbs in the opposite direction. In this way they use a selective tidal transport mechanism which permits them to exploit those tidal currents which take them in the direction of the spawning ground, probably with the aid of internal biological clockwork of tidal periodicity (Gibson, 1973; Gibson *et al.*, 1978; Metcalfe *et al.*, 2006). They cling to the sea bed when reversed tidal currents would tend to take them in the opposite direction.

Many other commercially important species have also been shown to complete their life cycles by migrating in one direction as planktonic larvae and in the opposite direction as adults. Typical among these are prawns of the genus *Penaeus* and their relatives, many of which live abundantly in the open seas of tropical and subtropical regions of the world. They occur in large populations that are sustained by recruitment of juveniles that migrate offshore from nursery grounds in estuaries and coastal lagoons to join the parent stock in the open sea. As they mature they reproduce and in their turn release planktonic larvae which continue the life cycle by making the return journey from the open sea to inshore nursery grounds again. Nowhere is this more evident than off the coasts of western Mexico (Edwards, 1978) where fishermen exploit large populations of adult prawns by trawling along the continental shelf, delivering their catches to processing plants on the coast where the prawns are shelled, frozen and packed for home and export consumption. In doing so, the open-sea fishermen compete with land-based cooperatives which fish for juvenile prawns of the same species, using fixed nets in adjacent coastal localities. That the two groups of fishermen were in fact competing for the same resource became evident when it was recognized that larvae released by the adult breeding stock on the continental shelf are somehow transported inshore by ocean currents, moulting and growing as they do so, until they reach the coastal nursery grounds. So, as they mature and leave the nursery grounds for the open sea the prawns are first subjected to the attentions of the inshore, land-based fishermen, with those that successfully join the adult population offshore, being subjected to further intense exploitation by commercial trawlers.

Within the vast lagoons of western Mexico juvenile prawns of various species are to be found, growing fast, at the rate of a millimetre each day, before heading seawards again as they begin to reach maturity. To reach the sea the migrating juveniles have to negotiate narrow exit channels from the lagoons into the mouths of river estuaries, and it is in these narrow *esteros* (Plate 17) that they risk capture by the shore-based fishermen. Prawn traps (*tapos*) (Plate 18) with fine mesh screens are set out across the full width of the esteros enabling capture of large quantities of prawns before they are able to return to the sea to spawn. Management of the entire fishery, in its two locations, is inevitably a complicated and diplomatic process, necessitating extensive study of prawn biology and lagoon ecology. Not least, fishery managers recognize the need to encourage healthy recruitment of late stage larvae along the esteros and into the lagoons. Necessary requirements are the lifting of tapos screens and the supervision of dredging operations to keep the esteros open at the times of greatest abundance of larvae when they arrive from the open sea.

After the larvae are spawned into the plankton above the offshore fishing grounds, they first make their way towards the Mexican coast on ocean currents. How this is achieved is not clear, but related studies on the larvae of pink shrimp, *Farfantepenaeus duorarum*, off Florida, USA, suggest that their daily rhythm of vertical migration in the plankton permits them to exploit the solar component of tides to carry them inshore at a speed of over 2 km each day (Criales *et al.*, 2005). Late in development, as the larvae enter estuaries of eastern Mexico or Florida, they tend to sink towards the bottom, swimming up from the bottom to exploit the inflow of seawater on rising tides, sinking back again as the tides ebb, making their way upstream into a lagoon where metamorphosis to the juvenile stage takes place. Whether or not penaeid prawns possess daily and tidal body clocks remains to be established, but they certainly behave as if they do.

There are, of course, even more spectacular examples of ocean wandering by species that may disperse for thousands of kilometres in the ocean before returning to specific breeding sites at particular times of year, often, as noted in Chapter 4, determined by spectacular feats of navigation. Trans-ocean migrations occur in large fish such as tuna, in marine turtles and in whales, not to mention aerial wanderers such as albatross and the Wideawake Tern. For these animals, as for the zooplankton species discussed at length in this chapter, there are clear practical difficulties in determining unequivocally that their rhythmic behaviour patterns are controlled by internal biological clocks. Convincing evidence for the possession of a biological-clock mechanism is usually demonstrated only by maintaining animals in constant conditions in the laboratory. However, by analogy with coastal species that are amenable to laboratory study, it is not unreasonable to assume that endogenous biological clocks are widespread in marine animals. Many ocean drifters certainly appear to behave as living clocks in response to cyclical changes in the environment and therefore seem likely to possess endogenous biological clockwork that controls their rhythmic behaviour.

*   *   *   *

Early studies by marine biologists tended to emphasize the over-production of eggs and planktonic larvae by organisms such as shore barnacles and crabs, on the assumption that eggs and larvae are randomly distributed in the sea. The production of vast numbers of offspring was viewed as being beneficial for the spread and survival of the species, but necessarily wasteful in view of the random nature of the process. From the evidence now available it is certain that if the release of eggs and larvae were truly random, then marine animals that produce small, swimming larvae would have to put even more energy into the reproductive process than is presently the case. Directional transport of free-swimming larvae that behave as biological clocks is clearly of common occurrence. Moreover, for shore-living species such as mussels, barnacles and crabs, other behavioural mechanisms that favour return of larvae to adult settlement sites are also being discovered. For example, as referred to in an earlier chapter, it has been demonstrated that late larvae of crabs returning from the open ocean to settle on New Zealand shores were, in the laboratory, attracted to recordings of the natural sounds of water movements on coastal reefs (Jeffs et al., 2003). Moreover, other studies in Chilean coastal waters and elsewhere suggest that tidal bores may play a significant role in transporting late larvae to their settlement sites inshore (Pineda, 1994; Shanks, 1995; Vargas et al., 2004). Tidal bores are most impressively seen in rapidly narrowing estuaries where the tidal range is substantial, as in the Bay of Fundy in North America and in the estuary of the River Severn in south west Britain. Rapid narrowing of the Bristol Channel and a tidal range of up to 10 m can produce a wall of water 2.5 m high that sweeps as roaring surf over sand flats up the River Severn at 25 km h^{-1}, most spectacularly during equinoctial high spring tides. Based on the theory of tides, undersea tidal bores should be widespread phenomena in many coastal regions throughout the world and could therefore play an important role in the inshore transport of the larvae of coastal species worldwide. The behaviour of larvae under such tidal conditions is less well understood than the vertical migration rhythms of daily and tidal periodicity that occur in open sea plankton species. However, present understanding of the behaviour of plankton is such as to permit a challenge to be made to physical oceanographers who suggest that randomly drifting plankton species can be used as inert indicators of the flow of ocean currents. It is clear that the behavioural characteristics of marine plankton animals should be quantified and factored into mathematical models of ocean current flow, which previously have sometimes assumed that zooplankton adults and larvae behave as passive particles (see Siegal *et al.*, 2003). Closer investigation of the clock-based rhythmic behaviour of planktonic adults and larvae, in relation to knowledge of ocean currents, should lead to improved understanding of the location of spawning sites and direction of migration routes of animals such as crustaceans and fish that are of commercial importance. Here, basic science is providing improved understanding for sustainable management of fishery resources, whilst at the same time posing new questions concerning the nature of the biological clock. Finally, then, we have reached the stage of suggesting that biological clock phenomena are so ubiquitous in marine organisms that indications are now required as to how those biological clock phenomena are controlled from within an organism's body. If biological rhythms are not solely driven by external geophysical variables, as Frank Brown originally suggested that they were, then one must address the questions as to where internal biological clockwork might reside and how it might function, tasks for the next and final chapter.

10

Living clockwork

The possession of a time sense allows an organism to set in motion at the proper time, processes whose final stage should be synchronized with set phases of a rhythmical environment

Janet Harker, 1964

Ever since publication of the challenging scientific article entitled *Biological clock in the unicorn* (Cole, 1957), research concerning the rhythmic behaviour and physiology of plants and animals has overwhelmingly supported the view that such processes are not controlled solely by cyclical changes in the environment, however subtle some of those variables might be. It has long been apparent that biorhythms can be expressed spontaneously, at periodicities somewhat different from those of geophysical variables, in isolation from the environment, leading to the conclusion that some form of internal biological clockwork is involved in their timing. It has been questioned as to how this free-running ability has provided adaptive advantage during the evolutionary process (Winfree, 1987) but, as discussed in Chapter 3, it seems to follow logically if, as appears to be the case, endogenous rhythmicity provides an organism with anticipatory capability in relation to potentially adverse or favourable phases of their cyclical environment. For example, for photosynthetic unicellular organisms early in evolution, anticipatory circadian physiology would permit the capture of the first photons at dawn (Suzuki and Johnson, 2001), would permit favourable phasing of incompatible processes such as photosynthesis and nitrogen fixation (Mitsui *et al.*, 1986) and permit favourable phasing of cellular events that are inhibited by sunlight (Pittendrigh, 1993). Similar advantages would accrue for higher plants and, for animals, there would be additional advantages in being able to anticipate adverse environmental conditions and the behaviour of predators or prey. For marine animals of tidal shores rhythmic anticipatory behaviour would seem to be particularly advantageous in reducing the risks not only of predation but also of desiccation at low tide. Whether such anticipatory behaviour enhances genetic fitness and survivorship has not been demonstrated experimentally, except in some prokaryote cyanobacteria cultures (Kando *et al.*, 1994) (see Chapter 3). However, because of the universal occurrence of biorhythmicity, which can reasonably be assumed to convey adaptive advantage to the organisms concerned, it is necessary to ask where the site of such biological clockwork is located and how might it function.

Since the process of evolution has repeatedly solved adaptive problems in different ways, such as for example in the diversity throughout the animal kingdom of mechanisms of flight, there is no a priori reason to suppose that all animals and plants have the same form of biological clock. On the other hand, since the phenomenon of endogenous biorhythmicity is apparent in a wide range of living things from those as complex as human beings to simple single-celled organisms, including prokaryotes, adaptation to the rhythmic environment may be a fundamental feature of life itself. Accordingly, at the outset, a search for the site of biological clockwork could not overlook the possibility of involvement of time-keeping at the level of components of living cells, even at the level of individual genes. This has been the approach of chronobiologists in recent years, though their efforts have focussed mostly on the control of biological rhythms in non-marine organisms and particularly those expressing circadian rather than circatidal rhythmicity.

Notable studies directed towards the search for a circadian clock in marine organisms have been carried out on the unicellular marine alga *Acetabularia*. This highly unusual organism consists of a single, huge cell, 3–5 cm long and about 1 cm in diameter with a 'rhizoid' attachment to the substrate. At one stage of development the nucleus of *Acetabularia* is located near the point of attachment, from where it can easily be removed and transferred to another plant. Accordingly, two cultures of this remarkable organism were initially entrained in 12 : 12 h light/dark regimes with opposite phases of the light and darkness. Then nuclei were reciprocally transplanted between plants of the two cultures and the rhythm of photosynthesis of each then recorded in continuous light. The rationale and outcome of the experiment are illustrated in Figure 10.1, in which it can be seen that the phase of the circadian rhythm of photosynthesis in each cell corresponds to the entrained phase of the implanted nucleus; peak oxygen consumption occurs during the expected dark phase of the host cell in each case (Schweiger and Schweiger, 1977).

LD 12:12------►LL

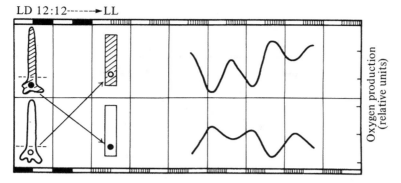

Figure 10.1 Experimental manipulation of the circadian rhythm of photosynthesis in *Acetabularia*. Two cultures were initially kept under L : D 12 : 12 h cycles with opposite phases of light and dark, after which nuclei from plants in each culture were reciprocally transplanted. After transplantation the rhythm of photosynthesis of each recipient cell (measured by oxygen production) was monitored in continuous light. Shaded bars, expected dark periods of the recipient cells (after Kluge, 1982).

However, whilst it was tempting to consider from these experiments that the basic circadian oscillator is located in the nucleus, earlier experiments with enucleated *Acetabularia* cells showed that they too exhibited a circadian rhythm of photosynthesis (Sweeney and Haxo, 1961). The most economical hypothesis at that stage was therefore to consider that circadian rhythmicity is intrinsic, not to the nucleus, but to the chloroplast as the site of photosynthesis, and that chloroplast rhythmicity is entrained by information from the nucleus. Temptingly, it was considered that a basic clock component resided in the chloroplast DNA, but that concept was discarded since it could also be shown that clock function persisted in *Acetabularia* in the presence of various antibiotics that were known to inhibit protein synthesis by the chloroplast (Sweeney *et al.*, 1967). Accordingly, since a choice of cellular clock hypotheses at that time was between a 'protein synthesis model' and a 'membrane action model' the latter was developed as a possible mechanism by which circadian rhythms could be generated, as will be discussed later.

Around the same time, other equally intensive studies were carried out on another unicellular marine alga, the dinoflagellate *Lingulodinium* (*Gonyaulax*) (Hastings and Sweeney, 1958; Sweeney 1969a, b, 1976), initially concerning its daily rhythm of bioluminescence. Phosphorescence in the sea is often caused by blooms of single-celled organisms such as *Lingulodinium* which continues with its nightly rhythm of spontaneous bioluminescence in cultures in the laboratory. The rhythm also continues for some time in constant darkness, indicating that it is truly endogenous, circadian, and under the control of internal clockwork. However, the amplitude of successive peaks quickly damps down in continuous darkness since the organism is photosynthetic and requires daily exposure to sunlight to fuel its metabolism. As with *Acetabularia*, early efforts to understand the basis of the circadian rhythm of luminescence in *Lingulodinium* focussed on the cell membrane since it was known that mechanical stimulation of luminescence was more effective during expected darkness than during expected daylight, even when the cells were maintained in continuous darkness (Christianson and Sweeney, 1972). Also, in another bioluminescent dinoflagellate, *Noctiluca*, it was found that mechanical stimulation of luminescence was preceded by an action potential measured at the cell surface, suggesting again that membrane function changes during the circadian cycle (Eckert, 1966). Here, however, it is necessary to consider the nature of biological clockwork, not only in unicellular organisms, but also in more complex organisms, such as animals whose evolutionary survival has depended upon adaptation to environmental cycles of tidal, as well as daily, periodicity. Such organisms are crabs and their relatives, which have been frequently referred to in this book so far, not least the now globally distributed common green shore crab *Carcinus maenas*. After initial characterization of clock-controlled tidal and daily walking rhythms in this crab (Naylor, 1958) many other similarly rhythmic aspects of its behaviour have subsequently been described (Naylor, 1988). Over many years, others too have studied various aspects of the biology of this species of crab and its relatives, to the extent that a sound understanding is available concerning crustacean lifestyles and physiology (see, for example, Bliss, 1982 onwards). This knowledge offered a useful basis from which to begin the search for the mechanism of crustacean biological clockwork, since circadian and circatidal periodicities were known to be evident in a number of physiological processes. In *Carcinus* they are expressed in the control of gill ventilation and oxygen consumption rhythms during respiration (Arudpragasam and Naylor, 1964), in oxygen consumption rhythms in *Emerita asiatica* (Chandrashakaran, 1965), in retinal pigment changes in two species of the hermit crab *Pagurus* (Ball, 1968) and the crayfish *Procambarus bouvieri* (Arechiga, 1977), in blood sugar concentrations in *Carcinus maenas* (Rajan *et al.*, 1979), in sensory and motor neuronal activity in *Procambarus bouvieri*, *Carcinus maenas* and *Nephrops norvegicus* (Arechiga *et al.*, 1974, 1980; Arechiga and Huberman, 1980), as well as in adult and larval locomotor activity (Naylor, 1988; Zeng and Naylor, 1996a, b, c). All of these physiological and behavioural activities, and others too, such as larval hatching (Zeng and Naylor, 1997) and juvenile moulting (Zeng et al., 1999), have been shown to persist rhythmically in constant conditions in the laboratory as if they were under the control of internal biological clocks. Within any one animal some of these rhythmic processes such as oxygen consumption, blood sugar changes, neural activity and locomotion may be functionally interrelated. However, whether or not the processes are interdependent, this kind of physiological symphony has provided ample stimulation for curiosity-driven studies to search for the underlying biological clock system that seems to control the timing of these rhythms. Such research has been carried out not only in crabs, but also in other crustaceans and more widely in coastal animals whose physiology is in tune with their rhythmic environment (see for example Strumwasser, 1965; Jacklett, 1982; Harris and Morgan, 1984c; Naylor, 1985, 1988).

Before continuing on the quest for the clock in invertebrates such as crustaceans, however, it is important to be reminded of some fundamental characteristics of endogenous biorhythms that must be explicable in the context of any clockwork mechanism that is proposed. An endogenous rhythm is a naturally occurring biological rhythm that continues to be expressed in constant conditions in the laboratory. Removed from the influence of environmental synchronizers, the pattern of the endogenous rhythms does not exactly match the external changes which recur slavishly in nature. In constant conditions endogenous rhythms only approximate to external periodicities and are therefore described as circatidal, circadian, and so on. Their periodicities may be shorter than those in nature, but are usually slightly longer, resulting in delayed drift of peak values of the function being recorded, when compared with the pattern in nature. In addition, as endogenous rhythms free-run in the laboratory, successive peaks become progressively less distinct, subsiding and spreading out when compared with their predecessors. Critically, however, the original pattern can be reinstated by appropriate treatment in the laboratory. In the shore crab, and some other marine organisms that have been tested, a short exposure to low temperature will re-start a rhythm of tidal periodicity in animals that appear to have lost their rhythmicity after several days in constant conditions in the laboratory (Naylor, 1963; Gibson, 1967), a response in Carcinus which is correlated with a dramatic release of hyperglycaemic hormone (CHH) (Chung and Webster, 2006). Furthermore, endogenous rhythms are generally considered to be expressed in an undisturbed pattern irrespective of the temperature to which an animal is exposed within its normal thermal range (Bűnning, 1967; Brady, 1979, 1982). This is an unusual characteristic of the physiological processes of cold-blooded animals, which normally speed up or slow down according to the rise and fall of temperatures to which the animals are exposed. In other words, biological clocks do not only occur in warm-blooded animals such as mammals and birds which are able to maintain a more or less constant body temperature. They also occur in cold-blooded animals such as crabs and other invertebrates which, by definition, do not normally regulate their body temperature. In fact, the physiological and behavioural rhythms of shore crabs do show transient responses to sudden rises and falls of temperature, but the responses are short-lived, suggesting that the crabs have some means of compensating for changes in temperature, at least in so far as biological clock function is concerned (Naylor, 1963). Indeed, as a generalization, it can be said that persistent endogenous rhythms of crabs and other organisms that have been studied, whether warm-blooded or cold-blooded, are functionally temperature-independent.

Initial attempts to suggest how crustacean physiological clockwork might function took into account the properties of endogenous rhythms and proposed that the most likely model was one of multiple clock control rather than control by a single clock (Figure 2.11). Away from environmental influences, a single biological clock might be expected to continue uninterrupted, run slow or fast or stop, none of which satisfactorily explains the behaviour of animals observed in constant conditions in the laboratory. On the other hand, multiple clocks, each approximating, say, to the tidal or daily periodicity, would gradually drift out of phase with each other and best explain the gradual subsidence and spreading of peaks of behavioural activity in constant conditions, as shown in Figure 2.11. Furthermore, that figure also shows how a low temperature shock and perhaps other stimuli such as vibration can easily be envisaged as a sudden resynchronizing stimulus which temporarily stops all the clocks, timing their restart in phase again to reinstate the original pattern of behavioural rhythmicity which is observed after the shock treatment.

But the question remains as to how the biological clock system of a cold-blooded animal could maintain its timing despite being exposed to increases or decreases in temperature. For shore crabs it was envisaged that this so-called temperature-independence could be achieved by postulating control by two physiological processes, one stimulating activity and the other inhibiting it (Naylor, 1963). A temperature rise that might cause clocks to run faster by stimulating a promoter system would be countervailed by over-stimulation of the inhibitor system, and vice versa. With such ideas in the background, the first attempts to characterize the clockwork of relatively advanced organisms such as crustaceans began by searching for hormonal and nervous processes that might promote or inhibit rhythmic behaviour. This approach coincided with a surge of interest in the processes of physiological control of moulting, growth and reproduction of crustaceans, many of which were of importance in the developing field of aquaculture. The scientific approaches in these fields of study tended to be similar, involving the removal of tissue or injection of tissue extracts, so-called 'classical endocrinology', in attempts to local-ize sites in the body where, for example, hormones controlling behaviour might be produced or released (Carlisle and Knowles, 1959).

Initial studies focussed on the X-organ sinus gland complex (Plate 19), situated in the prominent stalks of the eyes that are a feature of crabs, lobsters and crayfish. Various nerve cells within the X-organ syn-thesize hormones that are transported to the sinus glands (see Figure 10.4), where they are released into the haemolymph. It was previously known that some of these neurosecretory cells influenced the sequence of moulting and therefore the pattern of growth of crustaceans, and that other cells in the same organ influenced daily patterns of colour change (Carlisle and Knowles, 1959). Therefore, as a first step, it seemed logical to investigate whether other hormonal secretions from the X-organ might be involved in controlling the locomotor activity patterns of crabs in rela-tion to day and night and to tides. The first experiments with *Carcinus* involved surgical removal of the X-organ from crabs and recording their patterns of walking behaviour thereafter (Naylor and Williams, 1968). After full recovery from the operation the crabs were contin-uously active in the recording chambers, showing no signs of tidally rhythmic increases and decreases of walking activity that they normally showed on being brought freshly into the laboratory from the shores outside the laboratory. It was possible, since the crabs were given a recovery period after surgery, that their rhythmic behaviour would have been lost anyway. Certainly, unoperated freshly collected crabs even-tually lost their rhythmic pattern of behaviour when kept in constant conditions for some days (Naylor, 1958). However, though normally when freshly caught crabs were brought into the laboratory and kept for several days they could be induced to restart their tidal rhythms by cooling them at 4°C for a few hours (Naylor, 1963), this was not pos-sible in crabs from which the X-organ was surgically removed (Naylor and Williams, 1968). Crabs lacking X-organ cells were continuously hyperactive, even after full recovery from the operation. This important finding therefore suggested that the X-organ might be part of the crab's clockwork, perhaps by the release of a hormone that suppressed walking activity at particular times. It was hypothesized that the crab produces an inhibitor substance that was released rhythmically from cells of the X-organ and which induced quiescence at the times of low tide.

To test that hypothesis the next step was to make an extract of X-organ tissue from normal crabs, inject it into crabs from which X-organ tissue had been removed, and assess whether it would induce quiescence in hyperactive crabs. This turned out to be the case; injection of extract from eyestalk tissue suppressed walking activity in crabs but injections of muscle extract or sterile seawater into crabs maintained in control experiments did not (Naylor and Williams, 1968). Furthermore, in later experiments (Naylor *et al.*, 1971) it was found that X-organ extracts taken from crabs at low tide, when they were normally inactive, were more effective tranquillizers than extracts taken at the times of high tide. This suggested even more convincingly that the X-organ was functioning as at least part of the biological clockwork of the crab, by releasing an inhibitory agent into the crab's haemolymph at times of low tide when crabs are normally hidden beneath stones and boulders on the beach. Questions which could now be asked were whether crabs possessed a circatidal, blood-borne hormonal inhibitor component in their biological clockwork and, if so, how it might work.

Following related studies at the time which demonstrated spontaneous rhythmic activity in crayfish nervous systems, recorded using micro-electrodes coupled to an oscilloscope (Arechiga and Wiersma, 1969), similar experiments were carried out which established that daily rhythms of electrical output could also be recorded in the nervous system of the crab *Carcinus maenas* (Arechiga *et al.*, 1974). After locating in the optic peduncle 'activity neurons' that normally show increased firing rates when a crab is moving, recordings from such neurons were made when a crab was restrained. Under those circumstances, electrical recordings from the activity neurons were found to vary rhythmically and quite spontaneously over a 24-h cycle. Neuro-electrical discharges increased spontaneously during night-time and fell back to baseline levels during daytime, even in crabs kept quiescent in a recording room under continuous light and at constant temperature (Arechiga *et al.*, 1974). Here then a circadian rhythm in the generation of nerve impulses was being recorded, the discharge pattern of which could be correlated with the circadian component of the rhythm of walking activity shown by freshly collected shore crabs kept in constant conditions in the laboratory (Figure 10.2). Was it possible that

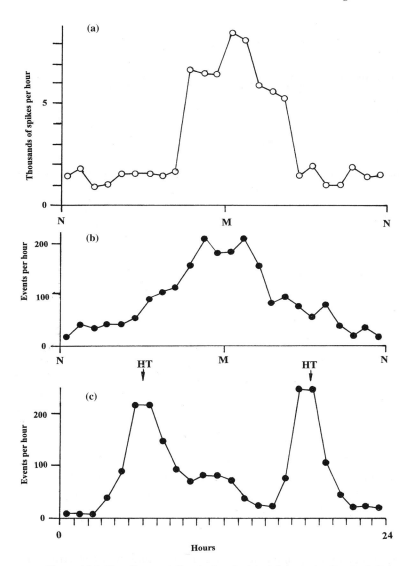

Figure 10.2 Circadian variation in the electrical firing rate of an 'activity' neuron of the shore crab *Carcinus maenas* (a), compared with hourly locomotor activity values of crabs collected each day throughout a semilunar cycle and recorded for 24 h in constant conditions. Locomotor data pooled around the times of midnight (M) in (b) and high tide (HT) in (c) (a, after Arechiga *et al.*, 1974; b and c, after Naylor, 1958; by permission of the Royal Society).

alongside the likely blood-borne hormonal inhibitor of circatidal periodicity that a neuro-electrical, circadian promoter component of crab clockwork had also been located?

If there were circatidal and circadian clock components in crab clockwork then the next logical step was to ascertain whether in some way they might interact. Since the hypothesized circatidal component had a tranquillizing effect on whole crabs, would it also suppress the electrical activity of nerves? Accordingly, extracts of X-organ tissues were injected into crabs whilst the electrical output of activity neurons continued to be monitored. The inhibitory effect of the extract on the whole crabs was indeed repeated in the output from recordings of their nerve action. Electrical output from nerves was partially suppressed when X-organ extract was injected into the haemolymph but not when muscle extract was injected as a control (Arechiga et al., 1974). So, it seemed that two possible physiological mechanisms could be proposed as being involved in controlling the rhythmic behaviour of crabs, prompting further enquiry into the nature of the inhibitory factor involved.

The next step then was to carry out biochemical separation of the extract obtained from X-organ cells to try to establish what the effective inhibitor might be. In these experiments the extract was first dialysed by passing it through a semi-permeable membrane, which allowed the small molecular fraction to pass through and which retained the larger molecule compounds. The injection experiments were then repeated, showing that the dialysed, large molecule fraction of the extract was not effective in suppressing nerve output. However, the dialysate that had passed through the membrane inhibited nerve activity exactly as the entire extract had done (Figure 10.3) (Arechiga et al., 1974). The effective substance was found not to lose its potency when heated to 80°C for five minutes, but to do so if incubated with pronase, an enzyme that digests proteins. On this evidence, the inhibitory factor from the X-organ was considered to be a peptide of small molecular weight. It was later named as neurodepressing hormone (NDH) (Arechiga et al., 1977) and found to be biochemically quite distinct from several other peptidic hormones secreted by cells in the crustacean eyestalk, such as crustacean hyperglycaemic hormone (CHH) which is involved in controlling blood sugar concentrations, pigment-dispersing hormone (PDH) and red pigment concentrating hormone (RPCH) which control eye pigment movements, and moult-inhibiting hormone (MIH) (Arechiga et al., 1985). The same substance has also been found in a number of other large crustaceans, including the Mexican freshwater crayfish *Procambarus bouvieri* (Huberman et al., 1979) and the Dublin Bay prawn *Nephrops norvegicus* (Arechiga

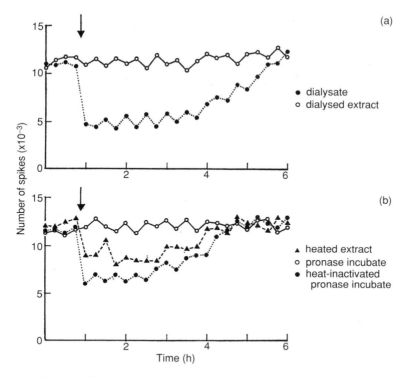

Figure 10.3 Spontaneous electrical activity in a *Carcinus maenas* tonic optomotor neuron, following injection into the crab of eyestalk extracts treated in different ways. Each point represents the number of spikes generated in a 15 min interval; arrows, times of injections. Injected (a) with dialysate or dialysed extract; and (b) with heated extract, pronase incubate or heat-inactivated pronase extract (after Arechiga et al., 1974; by permission of the Royal Society).

et al., 1979). Subsequently, Huberman (1996) used more sophisticated methods of analysis to ascertain that NDH is of even smaller molecular weight than initially determined. Its chemical nature has yet to be fully elucidated, but it remains as an inhibitory hormone of nerve action that is probably of universal occurrence in crabs, lobsters and crayfish.

Additional studies on the crayfish *Procambarus bouvieri* confirmed that greatest quantities of NDH are produced in cells of the X-organ which is situated on the median side of the medulla terminalis of the eyestalk (Figure 10.4) (Arechiga et al., 1985) and that in this freshwater species as well as some marine species of crustaceans NDH suppresses not only nerve action but locomotor activity too (see Naylor, 1988). When it was shown that injections of NDH suppressed locomotor activity in shore

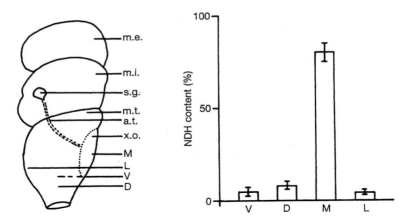

Figure 10.4 Neuro-depressing hormone (NDH) content in cell bodies isolated from different regions of the eyestalk of *Procambarus bouvieri* using the crayfish abdominal stretch receptor as the bioassay. m.e., medulla externa; m.i., medulla interna; s.g., sinus gland; m.t., medulla terminalis; a.t., axonal tract; x.o., X-organ; M, L, V and D, median, lateral, ventral and dorsal regions of the medulla terminalis (after Arechiga *et al.*, 1985; by permission of Springer Scientific and Business Media).

crabs (Figure 10.5) (Williams *et al.*, 1979a; Arechiga *et al.*, 1979), the circle was partly closed in relation to the point at which the clock-seeking experiments began with *Carcinus maenas*. However, a question still remained as to whether the site of production of NDH in the X-organ could be located. Among several types of cell to be found in the X-organ (Smith and Naylor, 1972), microscopic investigation revealed that one particular type of cell changed its appearance in phase with the tides. The cells appeared to manufacture and release their hormonal contents every 12 h or so, that is on a circatidal pattern (Williams *et al.*, 1979b). Moreover the cells continued to do this even when isolated in tissue culture in the laboratory, giving support to the notion that they may function as partially self-sustaining components of the crab clockwork. Originally it was speculated that the cells might be the tidally rhythmic source of NDH but more recent studies (Dirckson *et al.*, 1988; Wilcockson *et al.*, 2002) suggest that they produce only CHH and its precursor-related peptide, both of which are of larger molecular weight than NDH. Thus, whilst they may be partially involved in the control of physiological aspects of circatidal rhythmicity, some other cellular source of tidally rhythmic production of NDH still evades location. Nevertheless, a number of studies worldwide continue to search for nervous and hormonal mechanisms as a basis of the physiological clockwork of crustaceans and other

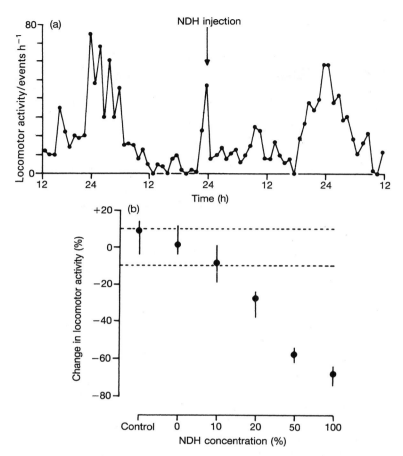

Figure 10.5 Effects of injecting NDH-containing fractions of eyestalk extracts on the circadian locomotor rhythm of *Carcinus maenas*: (a) hourly activity of one crab over a period of 72 h injected with 0.25 ml of partially purified extract (concentration index 100) at the time indicated by the arrow; (b) mean percentage change of locomotor activity in the 12 h period following injection of various concentrations of partially purified extract. The change in activity is determined from a comparison between the 12 h period after injection and the equivalent period 24 h earlier. Concentration is given as a percentage of the total protein content of the fraction, with 100% = 1890 g ml^{-1}. Circles and bars represent mean and range of at least three replicates; the horizontal dashed lines are ± 10% confidence limits of expected daily variation in activity (after Arechiga *et al.*, 1979; by permission of Elsevier).

marine organisms. Most agree that the circadian and circatidal rhythms in organisms such as the larger, stalk-eyed crustaceans are mediated via multiple physiological clock components, nerve structures in the eyestalk and in the supra-oesophageal ganglion (brain) being particularly implicated (Page and Larimer, 1975a, b; Larimer and Smith, 1980; DeCoursey, 1983; Arechiga *et al.*, 1985; Naylor, 1985, 1988). Interestingly, though, in a sessile-eyed crustacean, the estuarine amphipod *Corophium volutator*, the eye appears to have no role in controlling a circatidal rhythm of swimming (Harris and Morgan, 1984c). Using a precisely focussed chilling technique these authors found that rephasing of the amphipod's endogenous circatidal rhythm could be induced by cooling the brain (supra-oesophageal ganglion) and sub-oesophageal ganglion to $-3°C$ for 3 h. In contrast, rephasing was not induced by similarly cooling the telson or mid-body regions a very short distance from the suprao-esophageal ganglion (Figure 10.6). It was concluded that *Corophium* appears to possess two anatomically discrete pacemakers at the physiological level of its circatidal clockwork (Harris and Morgan, 1984c).

In marine molluscs that have been studied in this respect, the eyes and optic tract of the nervous system are again especially implicated in the control of circadian periodicity of the marine opisthobranch mollusc, the sea hare *Aplysia* (Jacklett, 1969, 1974, 1982). Indeed, a circadian rhythm of spontaneous compound action potential (CAP) output can be recorded continuously via a suction electrode attached to an isolated eye of *Aplysia* kept in tissue culture fluid (Figure 10.7). The amplitude of the rhythm peaks is proportional to the number of cells retained in the preparation and the circadian output can be stopped by adding sodium proprionate to the culture fluid, after which the CAPs continued firing arrhythmically. Each eye of *Aplysia* appears to function as an independent pacemaker capable of driving the circadian locomotor activity rhythm of this animal (Lickey and Wozniak, 1979). As in many crustacean clock systems, the eyes may be linked to other rhythmic physiological processes by neurosecretion, since it is known to be the source of several polypeptides (Jacklett, 1982). However, similar studies to those on *Aplysia* have been carried out on the closely related mollusc *Bursatella*, in which, analogous to *Corophium* amongst the crustacean examples discussed, the eyes appear to play only a minor role in the control of circadian rhythmicity; in *Bursatella* physiological clockwork appears to be primarily under the control of extra-ocular pacemakers (Block and Roberts, 1981).

Interplay between hormones and electrical activity in the nervous system also appears to be at the core of the circadian biological clock network system in various vertebrates that have been studied, namely

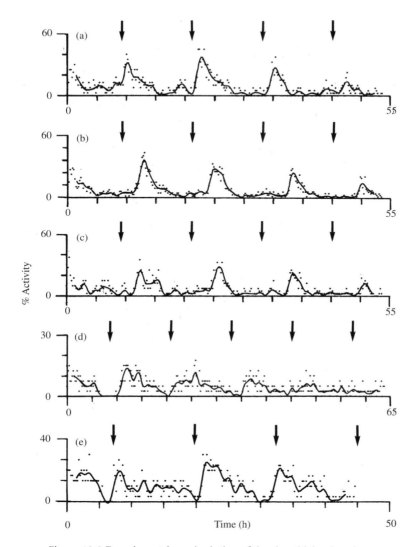

Figure 10.6 Experimental manipulation of the circatidal swimming rhythm of the amphipod crustacean *Corophium volutator*: (a) endogenous rhythm of the freshly collected specimens in constant conditions in the laboratory; (b) free-running rhythm of amphipods chilled to −3°C for 3 h before recording began; (c) free-running rhythm after localized chilling of supra-oesophageal ganglion; (d) free-running rhythm after localized chilling of sub-oesophageal ganglion; (e) free-running rhythm after localized chilling 1.5 mm posterior to the sub-oesophageal ganglion. Arrows, times of expected high tide (after Harris and Morgan, 1984c).

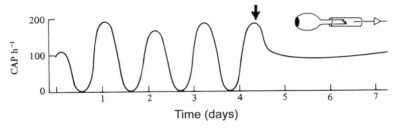

Figure 10.7 Circadian rhythm of compound action potential frequency
(CAP h^{-1}) recorded from the optic nerve of an isolated eye of the sea hare
Aplysia, maintained in organ culture in constant darkness. The eye is
shown in the inset with a recording suction electrode attached to the cut
end of the optic nerve. At the time indicated by the arrow, propionate was
added, which stopped or disconnected the clock, leaving CAP activity
continuing arrhythmically (after Jacklett, 1982).

fish (Cahill, 2002), birds (Brandstatter, 2002) and mammals (Menaker
and Tosini, 1996). In mammals the suprachiasmatic nuclei (SCN) in the
hypothalamus of the brain is regarded as a 'master clock' (Ralph *et al.*,
1990; Welsh *et al.*, 1995), much earlier studies having also shown that
circadian timing of drinking, locomotor activity, sleep and temperature
could be eliminated by isolation of that tissue (Stephan and Nunez, 1977).
This complex, of some 20 000 cells in rats, directly controls behavioural
and other circadian rhythms and, at the same time, regulates rhythmic
release of the hormone melatonin from the adjacent pineal gland. The
blood-borne melatonin then feeds back to influence the rate of electrical activity in the SCN itself. In birds, in contrast, the pineal gland itself
is considered to be the site of a self-sustaining circadian oscillator that
regulates the release of melatonin (Follett, 1982). In this regard, the role
of melatonin is of particular interest in chronobiology; it is an evolu-
tionarily conserved molecule occurring widely in living organisms from
unicells, through algae and other plants, to invertebrates, as well as vertebrates (Arnoult *et al.*, 1994; Hardeland *et al.*, 1995; van Tassel *et al.*, 2001;
Hardeland and Poeggeler, 2003; Pape and Lüning, 2006). Together with
the key enzyme regulating its biosynthesis it is involved in the transduc-
tion of photoperiodic information from the rhythmic environment to the
organism's physiology (Vivien-Roels *et al.*, 1984; Smith, 1990; Vivien-Roels
and Pevet, 1993; Itoh and Sumi, 1998; Withyachumnarnkul *et al.*, 1992).
Variously considered to have a role in affecting pigment cells and the photoperiodic control of seasonal reproductive patterns, in some organisms
it is also evidently involved in mediation of circadian behaviour. Certainly
the ingestion of medicinal melatonin is commonly thought to aid with

the phase-setting of human circadian rhythms induced by jet-lag, though the full implications of such use of melatonin require further elucidation (see Arendt *et al.*, 1995; Foster and Kreitzman, 2005). Among marine animals, several crustaceans have been studied in regard to the question as to whether melatonin is involved in biological clock control of locomotor and various other rhythms of circadian periodicity. For example, Tilden *et al.* (2003) demonstrated a correlation between increased haemolymph melatonin concentrations and peak circadian locomotor activity in the fiddler crab *Uca pugilator* (Plate 16), with earlier demonstration that experimentally administered melatonin at a particular phase of the day/night cycle could induce phase shifts in peak concentrations of glucose and lactate in the crab haemolymph (Tilden *et al.* 2001). In contrast, in the Norway lobster, *Nephrops norvegicus*, Aguzzi *et al.* (2009) showed that peak concentrations of haemolymph melatonin were not causally correlated with peaks in circadian locomotor activity rhythms. Experimental rephasing of the circadian rhythms of *Nephrops* was not followed by related changes in the temporal pattern of haemolymph melatonin, suggesting that melatonin was not involved in circadian clock control. However, this finding did not exclude the possibility that melatonin production is involved in the photoperiodic control of seasonal changes in growth and reproduction in this crustacean.

The anatomical sites and physiological mechanisms generating the various periodicities of biorhythms in marine animals, from tidal to annual, are still largely unexplored. They remain fertile fields of study, as also do the frequency-coupling mechanisms that occur between tidal and daily patterns of behaviour. There are however models, in the form of hypotheses, based upon experimental work so far, that lend themselves for future testing. Indeed, at the outset there are competing hypotheses on such a basic question as to how circatidal rhythms themselves are controlled. Intuitively, on the evidence presented so far, one is led to the notion that coastal animals which exhibit day-related and tide-related patterns of behaviour (Figure 10.2) possess internal clocks of circadian and circatidal periodicities. Notwithstanding the lure of that idea, Palmer (1995a, b, 1997) proposed a model to explain such behaviour in the form of the circalunidian clock hypothesis. The model is based on observations that the twice-daily tidal peaks of a tidal rhythm may behave differently from each other and that the two peaks may scan the solar day at different rates, each appearing to adopt their own period. Added to this, one peak might disappear, temporarily or permanently, while the other remained intact, and one peak might split while the other remained unchanged. These phenomena are observed in a number of marine animals recorded in constant conditions in the laboratory, leading to the conclusion that the two peaks in any one 24-h interval are acting quite independently of each other. By way of explanation, Palmer (1995a, b, 1997) postulated that each cycle of peaks is controlled by its own clock, and that the basic periodicity of each clock is approximately that of the lunar day, that is 24.8 h. The two independent clocks are considered to be strongly coupled at times when the expressed behavioural rhythms appear to be clearly tidal or circatidal, but become decoupled when there are differing rates of drift, splitting or disappearance of peaks. Earlier evidence suggesting that the rhythmic system of tidally rhythmic animals might have a fundamental period of about 25 h rather than 12.4 h was presented by Enright (1975). Under constant conditions the spontaneous rhythmicity of three species of sand-beach crustaceans appeared to repeat itself, not between successive high tides, but at about twice that interval. However, as Enright (1975) also pointed out, the animals were collected from a Californian shore where they would have experienced a complex tidal pattern with markedly unequal semi-diurnal tides (see Chapter 1), and their behaviour reflected that complexity. Some recent studies have supported the circalunidian hypothesis (see Thurman, 2004), but it has yet to be critically tested, and the alternative hypothesis of combined circadian and circatidal clockwork should also be considered, particularly since arguments in favour of true circatidal rhythmicity have been advanced (Neumann, 1981; Naylor, 1982). The two-clock circalunidian hypothesis explains circatidal rhythmicity in two intertidal invertebrates which appear not to have circadian rhythmicity (Williams, 1998), but the apparent lack of circadian rhythmicity has been shown to be reversed in another intertidal animal dependent upon its feeding state (Reid, 1988), so the phenomenon may be latent when not superficially apparent.

It is apparent that circadian timing ability is widespread among living organisms and it is reasonable to suppose that it occurs typically, alongside circatidal timing, however that is achieved, in intertidal organisms which have been exposed to tidal and daily variables over evolutionary timescales. After all, rhythmicity of circadian periodicity is characteristic of animals living just above and just below tidemarks, and such animals are often close relatives of intertidal forms. Also, it is the case that the shore crab *Carcinus maenas* exhibits both circadian and circatidal rhythmicity in summer (Figure 10.8) when it lives intertidally, but reverts primarily to circadian rhythmicity in winter when it resides predominantly below tidemarks (Naylor, 1962; Reid and Naylor, 1989). Moreover, in the case of *Carcinus* in winter it is the ability to express

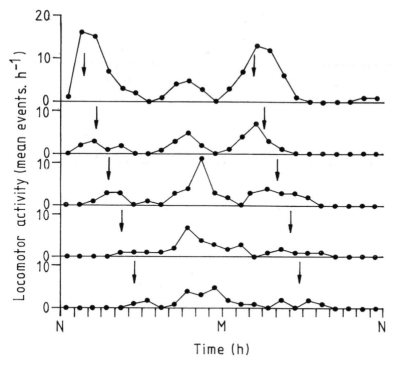

Figure 10.8 Locomotor activity of shore crabs *Carcinus maenas* kept for
5 days in constant conditions in the laboratory. Note increased activity at
each time of expected high tide (arrows) and around expected midnight
(M). N, expected noon (redrawn from Naylor, 1963).

circatidal rhythmicity that is retained in latent form, since it can be readily reinstated by a short session of immersion in dilute seawater (Reid
and Naylor, 1989). Convincingly, too, the timing of immersion in cold
water sets the timing of circatidal rhythmicity quite independently of
the crab's circadian timing, suggesting that it possesses both circadian
and circatidal clockwork (Reid and Naylor, 1989). Indeed, since marine
organisms pre-date their terrestrial relatives it has even been speculated
that circadian clocks might originally have evolved from tidal oscillators (Wilcockson and Zhang, 2008). In any case, alongside the circalunidian hypothesis, it seems reasonable to consider the notion of control
by circadian and circatidal clocks as an alternative hypothesis to explain
behavioural rhythmicity in at least some intertidal animals.

The circadian/circatidal hypothesis (Naylor, 1996, 1997), based on
studies particularly of shore crabs, proposes a model which incorporates a
circadian oscillator which is expressed at approximately 24 h periodicity

and a circatidal, approximately 12.4 h, oscillator running at a slightly longer relative periodicity, completing two oscillations in about 24.8 h. Deriving from physiological evidence presented earlier in this chapter, the circadian oscillator can be envisaged as an activity promoter driven by the circadian rhythm of neuro-electrical activity which interacts with a circatidal rhythm of release of an inhibitory hormone which partially suppresses nerve action and walking activity. For a hypothetical crab one can therefore run the model, allowing 6.2-h episodes of inhibition to progress at tidal (12.4-h) periodicity in relation to an assumed stationary 24-h rhythm of activity promotion. In so doing the two oscillators in the model influence each other such that the activity output by the hypothetical crab can be plotted and compared with the spontaneous activity rhythms of freshly collected crabs recorded in constant conditions in the laboratory. When the two sets of data are plotted together (Figure 10.9), it can be seen that the model-generated activity rhythm shows temporary disappearance of peaks, peak splitting, and varying drift rate of peaks, the entire plot more or less exactly matching the activity patterns of real crabs. All of the characteristics of crab rhythms, claimed to be explicable only by the circalunidian hypothesis (Palmer, 1995a), can therefore be explained by the alternative hypothesis of interacting circadian and circatidal oscillators. Interacting circadian and circatidal oscillators also offer a more readily explicable basis for circasemilunar rhythmicity by the 'beat' phenomenon, whereby peak circadian and circatidal rhythms coincide and enhance each other every 14–15 days. Further support for the circadian/circatidal hypothesis also arises because there is evidence that the proposed interacting promoter and inhibitory oscillators each appear to have a firm physiological basis. Moreover, the interaction of promoter and inhibitory rhythms could also help to facilitate the apparent temperature-independence of crab behavioural rhythms.

*   *   *   *

Aldrich (1997) concluded that 'the case rests without disproof' of either the circalunidian or circatidal/circadian hypotheses but, whichever of these two, or any other hypotheses, eventually prevails after future research, the underlying cellular and molecular timing mechanisms still remain to be elucidated. Clockwork in multicellular animals may be mediated physiologically through nervous and hormonal mechanisms, but circadian and tidal rhythms are also observed in single-celled organisms, as described in detail earlier in this chapter and in Chapter 2. In such unicellular organisms in particular, but in multicellular animals too, an early approach to characterize the properties of the cellular clock, determined by techniques available at the time, was to seek to change

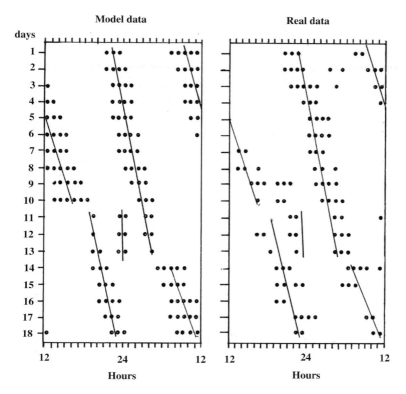

Figure 10.9 Model-predicted data and real data of locomotor activity of individual crabs *Carcinus maenas* freshly collected each day for 18 days and kept in constant conditions for 24 h. Dots are hourly activity values equal to or greater than the daily average. Trend lines in the real data are those predicted by the circadian/circatidal clock model (adapted from Naylor, 1996; 1997).

clock-timing by exposure to various inhibitory or stimulatory chemicals known to affect biological processes. The substances applied include ethanol, antibiotics such as cyclohexamide and valinomycin, heavy water, and many others. The treatments were carried out seeking to test two hypotheses in particular; one that cellular clock function may relate to cyclical synthesis of protein, the so-called chronon hypothesis (Ehret and Trucco, 1967) and the other that it may relate to cyclical transfer of substances across cell membranes (Njus *et al.*, 1974). As discussed earlier in this chapter, the 'membrane model' was favoured as a result of pioneering studies on the unicellular algae *Acetabularia* and *Lingulodinium* (*Gonyaulax*). It was also supported when it seemed that unicellular prokaryotic organisms such as cyanobacteria appeared not to exhibit circadian rhythmicity, evidently related to the fact that their nucleus and other cell bodies are not membrane bound as they are in eukaryotes. However, membrane action alone could not account for circadian rhythmicity when it was discovered that the prokaryotic cyanobacterium *Synechococcus* exhibits endogenous circadian rhythms of cell division (Sweeney and Borgese, 1989). Meantime, clock function had been shown to be disrupted by several of these chemical challenges in a number of unicellular and multicellular marine organisms, but findings show little consistency. Membrane active compounds such as ethanol and valinomycin induce shifts in the timing of circadian rhythms in the unicell *Lingulodinium* (*Gonyaulax*) (Sweeney, 1976) and the optic nerve of the sea hare *Aplysia californica* (Jacklett, 1982), and of the circatidal rhythms of the crustaceans *Excirolana chiltoni* (Enright, 1971) and *Corophium volutator* (Harris and Morgan, 1984b). In contrast, cyclohexamide, an antibiotic that inhibits DNA translation to protein, had little effect on the rhythmic patterns of behaviour of the isopod crustacean *Excirolana chiltoni* (Enright, 1971) and the amphipod crustacean *Corophium volutator* (Harris and Morgan, 1984b). Confusingly, cyclohexamide does lengthen the period of the rhythm of the unicell *Euglena* (Jacklett, 1982) and it inhibits the synthesis of neurodepressing hormone (NDH), which has a role in the control of large crustaceans such as crayfish and crabs (Arechiga *et al.*, 1985). However, these chemical challenges of clock function were blunt tools in studies from which the most likely outcome was perhaps always likely to be that membrane-action and protein-synthesis models were not mutually exclusive, thus supporting a coupled translation–membrane model proposed by Schweiger and Schweiger (1977). Both types of process would appear to be involved in crab clockwork that includes nerve membrane activity which stimulates body movements and synthesis of a hormone that inhibits nerve action and locomotor activity. But, even if these two types of process are involved in controlling animal rhythms, they too must also be driven by even more deep-seated timing processes.

*   *   *   *

It is apparent in studies of crabs, as it is in other animals too, that biological rhythmicity is an inherited characteristic of living things. Therefore, it soon became logical to conjecture that the fundamental biological clock is under genetic control and must reside at the molecular level of animal organization, a field of study ripe for investigation once appropriate techniques became available. These developed during the latter part of the twentieth century as powerful new tools with which to study the genetic makeup of plants and animals. The breakthrough came when

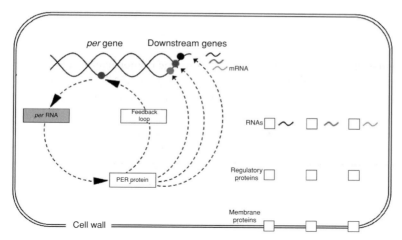

Figure 10.10 Early schematic molecular clock model (redrawn from Hardin *et al.*, 1990; Arechiga, 1993).

it was discovered how to cleave the DNA of an organism's genome by the use of specific enzymes in order to isolate individual genes. Genetic engineering was born when it also became possible to cut up DNA into known pieces, to re-assemble them in different ways, and even to transfer genes from one organism to another.

For circadian biological clock systems, early intuitive conclusions that living systems possessed inherited time-keeping ability were supported when genes known as period (*per*) genes were characterized in a number of animal species. Such, so-called clock genes, were first identified in *Drosophila melanogaster* by Konopka and Benzer (1971). They found mutant fruitflies whose rhythms were of longer or shorter period than the usual circadian pattern and back-crossed them with normal flies. The numbers of normal and abnormal offspring bred from the crosses occurred in such proportions as to conclude convincingly that the ability to express circadian rhythmicity was under gene, that is, *per* gene, control. That discovery led eventually to the formulation of early hypothetical molecular models that proposed how *per* genes might function as part of the basic biological clock mechanism. A schematic representation of such a model and how it might operate within a single cell was formulated by Hardin *et al.* (1990), based also on studies of the fruitfly *Drosophila*. That model was subsequently elaborated by Arechiga (1993) in the context of studies of biological rhythms in crustaceans, including marine species, as shown in Figure 10.10. The models postulated that in the genome of a cell, such as a nerve cell, a *per* gene first produces *per*

RNA and then PER protein as part of the normal process of transcription of RNA from a DNA template during protein synthesis. Importantly, it was found in the *Drosophila* investigation that expression of RNA occurred with circadian periodicity; the amounts of RNA measured were high in the morning and low at night. Then, since RNA is a vital component of the process of protein synthesis, rhythmicity of its production was envisaged as controlling the daily rhythm of production of PER protein which directly, or by influencing downstream genes, affects the synthesis of a host of regulatory and membrane proteins responsible for the expression of overt rhythms at the cellular level. Critically, too, for the molecular clock to function effectively, PER protein was thought in some way also to act as an inhibitor of its own RNA synthesis for part of the circadian cycle. A negative feedback loop such as this was, and still is, regarded as an essential feature of the generation of rhythmicity in the basic molecular biological clock mechanism, though its mode of action by PER protein alone was difficult to explain.

More recently, other clock genes and associated feedback loops have been identified in various animals in particular, such that even more complicated molecular clock models have been proposed. The additional clock genes include *tim* (timeless) (Sehgal *et al.*, 1995), *cycle* and *clock* (Vitaterna *et al.*, 1994). Based on those findings a refined version of the *Drosophila* circadian clock model, probably applicable to a wide range of animals, including marine species, can be formulated (Figure 10.11). On this model (Figure 10.11a) the transcriptional proteins of *clock* and *cycle*, designated CLK and CYC, bind together in the so-called E-box regions of DNA upstream of the clock genes *per* and *tim*, switching on their transcription (1). Then *per* and *tim* messenger RNAs accumulate in the cytoplasm throughout the daytime (2), peaking in the late afternoon, followed by translation into their equivalent proteins throughout the night (3). At that point the negative feedback component of the model (Figure 10.11b) comes into play when the PER and TIM proteins begin to enter into the nucleus and to degrade in the cytoplasm (4). As with PER protein alone, it is difficult to envisage how the PER/TIM protein complex alone could inhibit its own transcription by *per* and *tim* genes as the negative feedback component of the molecular clock. However, the PER/TIM complex of proteins evidently binds to the complex of proteins deriving from *clock* and *cycle* genes and it is this entire protein complex which now inhibits transcription of the *per* and *tim* genes (5). Then, in the absence of *per* and *tim* RNAs (6) and continuous degradation of PER and TIM proteins in the cytoplasm (7), input of PER and TIM proteins to the nucleus ceases, the CLOCK and CYCLE transcriptional proteins in the absence of PER and

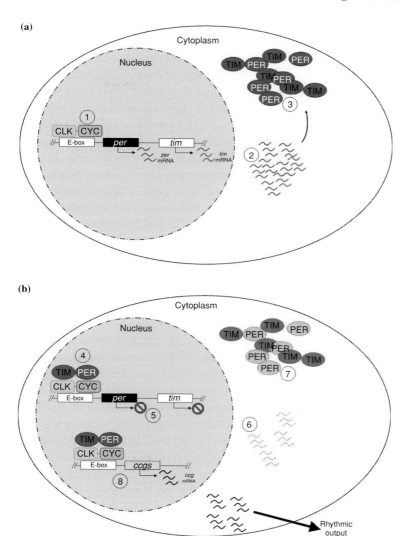

Figure 10.11 Generalized model of genes involved in circadian clockwork, developed mainly from studies of *Drosophila*, but with potential wider application including to marine animals (see text for explanation; after D. Wilcockson, unpublished).

TIM proteins revert to switching on the *per* and *tim* genes, and the loop restarts. Hence the system generates a circadian pattern of production of clock proteins which regulate the transcription of clock controlled genes (*ccgs*) (8), which induce control of rhythmic outputs at the physiological and behavioural level of the organism concerned.

Discovery of a suite of clock genes not only helped to clarify how a negative feedback loop might operate at the molecular level, but it also suggested one possible way in which entrainment of a circadian rhythm might take place at the molecular level of biological organization. When the *tim* gene was first identified (Sehgal *et al.*, 1995), it was discovered that its protein products, though stable in darkness, were rapidly degraded when exposed to light, suggesting quite strongly a possible molecular mechanism by which *Drosophila* circadian rhythms are cued by the environmental cycle of day and night. Subsequently, a further gene, thought to be involved specifically in the entrainment process, was discovered in *Drosophila*. Named *cry* (cryptochrome) (Hall, 2003), it appears to act in the cue-setting process by influencing degradation of protein products of the *tim* gene. Paradoxically though, mutant *Drosophila* without the *cry* gene are still able to entrain their circadian rhythms to the day/night cycle, indicating that the full story of clock-setting in *Drosophila* has yet to be revealed.

Suzuki and Johnson (2001) comment that circadian control permeates all levels of biological organization, drawing particular attention to pioneering work on the circadian basis of cell division, physiology, biochemistry and genetics in the marine unicellular algae *Lingulodinium* and *Acetabularia* and the cyanobacterium *Synechococcus*. They note that in the prokaryote *Synechococcus elongatus* a cluster of three *kai* genes has been identified, rhythmic expression of which is reminiscent of clock genes in eukaryotes such as *Drosophila*. Deletion of any or all of the three genes eliminates circadian rhythmicity without affecting viability. Moreover, feedback loops include a histidine protein kinase, *sosa*, disruption of which results in arrhythmicity of expression of most *kai* genes in moderate to bright light (Iwasaki *et al.*, 2000). At all levels of biological organization, therefore, clock function appears to use the same fundamental mechanism, namely an auto-regulative, negative feedback loop involving a number of genes. However, in prokaryotic cyanobacteria, perhaps even more than in eukaryotes, the totality of the clockwork remains tantalizingly veiled (Suzuki and Johnson, 2001).

Notwithstanding the unknowns, modern models of living clockwork also postulate that, from the basic molecular clock, rhythmic signals are probably transmitted to other, so-called 'output genes'. These would control rhythmic production of additional RNA and protein downstream in the regulatory process, en route to the appearance of overt rhythms at the cellular level. In unicellular marine organisms such as the planktonic dinoflagellate, *Lingulodinium* (*Gonyaulax*), referred to earlier, one such cellular rhythm generated in this way would be its circadian rhythm of bioluminescence (Hastings and Sweeney, 1958; Sweeney, 1969a). Light emitted by these single-celled organisms is generated on their body surfaces by oxidation of the protein luciferin which is synthesized within the cell body. The components of the biochemical reaction, the enzyme luciferase, the substrate luciferin and the substrate binding protein LBP, all oscillate in concert during the circadian cycle (Johnson *et al.*, 1984; Morse *et al.*, 1990), suggesting that they are under the control of a single oscillator. However, even in a single-celled organism such as *Lingulodinium* there may be more than one pacemaker. In this unicellular alga the circadian rhythms of bioluminescence and aggregation have been shown to run independently, demonstrating that they must be controlled by separate oscillators (Roenneberg and Morse, 1993). The aggregation rhythm, referred to in Chapter 7, is related to photosynthesis which would be more efficient by aggregation of the cells at the sea surface by day, behaviour which is comparable with that of vertical migration in zooplankton.

Developing the models further, cycles of synthesis of other proteins can be envisaged as occurring within individual cells of more complex, multicellular organisms, such as shore crabs and other marine animals. From there it is an easy step to suggest that such cellular clocks then control the rhythmic expression of nerve action and cyclic production of enzymes and hormones. These would then lead directly to the appearance of rhythmic physiology and behaviour at the level, say, of the whole crab on the beach. Therefore, on such a molecular model of biological clockwork, genes such as *per* and *tim* and their feedback loops can be considered to function as the primary pacemaker system that regulates, via secondary cellular oscillators, the drives and inhibitions generated by nerve and hormone action, to shape the eventual expression of biorhythms of physiology and behaviour in the whole organism. Even so, it appears to be the case that the output of the molecular clock system, rather than its entrainment process, is the least understood aspect of circadian rhythmicity (Williams and Sehgal, 2001).

*   *   *   *

Thus far molecular clock models have been developed mostly as if they were controlling circadian systems that drive daily rhythms of behaviour. It is still an open question as to whether coastal animals that exhibit tidal rhythms possess clock genes that express at circatidal periodicity in addition to clock genes of circadian periodicity. On the evidence available so far, it would be surprising if in coastal animals natural selection had not favoured genes, or at least the secondary cellular oscillator systems which they control, that express at approximately tidal *and* at approximately daily periodicity. Molecular approaches to the study of circatidal clockwork are still in their infancy but have been applied in a range of intertidal-living animals, including polychaete worms and crustaceans. For example, insect canonical clock gene orthologues that can be traced back to a common ancestor have been identified in the sand-beach isopod crustacean *Eurydice pulchra* (C. P. Kyriacou, pers. comm.), the rhythmic biology of which has been described in earlier chapters.

Preliminary evidence is also available for tidally related cycles of transcribed RNAs from many genes in the heads of *Eurydice*, but specific associations between individual clock genes and RNA transcripts have yet to be characterized. Functional links have also to be established between clock genes, output genes and tidally varying release of hormones into the blood circulatory system. One hormone which is a target for study in *Eurydice* is pigment dispersing hormone (PDH). This is a commonly occurring blood-borne hormone in crustaceans, rhythmic expression of which is involved in the control of daily changes of body colour. In *Eurydice pulchra* the beautiful arrays of black pigment cells on the body, from which it derives its specific name, disperse and concentrate their contents rhythmically in a circadian pattern. In continuous dim light the isopods are pale during expected night and dark during expected day. Rhythmic expression of hormones such as PDH in *Eurydice* and NDH in *Carcinus* has yet to be causally linked to cyclical clock genes, but it is reasonable to propose a working hypothesis that this is the case.

Given the acknowledged widespread occurrence of clock genes in living organisms, the molecular clock model illustrated in Figure 10.11 could have general applications and, with the addition of the nerve and hormonal mechanisms discussed earlier, could be hypothesized for coastal animals such as the shore crab *Carcinus maenas*. For such intertidal animals, which exhibit circadian *and* circatidal rhythms, a simple conjectural model of their biological clockwork might postulate fundamental control by clock genes that express at circadian *and* at circatidal periodicities. In the case of the shore crab, a circadian clock gene system can be envisaged as controlling the circadian rhythm of neuro-electrical activity which serves as a promoter of walking activity. Added to that, a clock gene system of circatidal periodicity can be envisaged as a control mechanism for the circatidal rhythm of production of neuro-depressing hormone that serves as a periodic inhibitor of walking activity. However, as speculative as this model is, the mechanism whereby phase-setting takes place for tidal clockwork is also poorly understood.

Circadian and photoperiodic timing mechanisms were first described in photosynthetic organisms, dependent as they are upon sunlight. Adaptation to daily and seasonal fluctuations in sunlight must have generated strong selective pressure not only for the evolution of circadian phase-setting *per se*, but also as a mechanism for the measurement of seasonal changes in day length (Johnson, 2001). The mechanisms are based on the occurrence in plants of two, interconverting forms of the pigment phytochrome. One form, which absorbs red light, is biologically active in the stimulation of enzyme reactions, whilst the other, which absorbs far-red light, is biologically inert. During daylight, which has an excess of red light, the red light absorbing form builds up in the plant, only to convert to the far-red absorbing form at night. As a result, phytochrome provides a plant with the ability to detect day and night as the input signal for its circadian phase-setting, changes in day length being similarly monitored as the basis for its photoperiodic time measurement (Vince-Prue, 1982; Thomas and Vince Prue, 1997). Among invertebrate and vertebrate animals, too, at the physiological level, the role of light in entraining circadian rhythms and photoperiodic time-measurement is quite well understood (Page and Larimer, 1972; Saunders, 1982; Follett, 1982; Arechiga, 1993). Moreover, at the molecular level, it is known that in the fruit fly *Drosophila* protein products of the clock gene *tim* are degradable by light, probably aided by the involvement of entrainment genes such as *cry*, suggesting how light cycles may directly synchronize the circadian clock system (Sehgal *et al.*, 1995; Williams and Sehgal, 2001). However, the genetic basis of synchronization by tidal variables has yet to be investigated. Clock-setting of circatidal rhythms in coastal animals must occur via the physiological processes that perceive cyclical changes of environmental temperature, hydrostatic pressure, wave turbulence and salinity (see Chapter 3), but it remains to elucidate whether synchronization with the environment occurs by entrainment of intermediate oscillating physiological processes or of the basic pacemaker system itself. Moreover the role of genes such as *cry* in the entrainment process of circatidal rhythms is, as yet, unknown. However, because of the universal nature of the genetic code whereby the same gene-derived proteins are produced in organisms from bacteria to humans, molecular clockwork proposed after extensive study of one organism can reasonably form the basis of hypotheses put forward for testing in other organisms. This is a scientific procedure sometimes called 'model-hopping'.

Whether the hypothetical clock model proposed above eventually proves to be correct for marine animals such as *Carcinus* it is nevertheless based upon a number of sound observations, tested hypotheses and justifiable model-hopping, with potential development as a theory of relevance to other marine animals too. However, progress towards firmer theory concerning the molecular basis of living clockwork in marine animals awaits further study. Do coastal animals possess both circatidal and circadian clock genes? Alternatively do they possess only circadian clock genes or possibly circalunidian clock genes, some of which control tidally related behaviour by cycling out of phase with each other? Do cells of shore crabs, or any other coastal animals that show circatidal and circadian rhythms, express circadian and circatidal cycles of clock gene RNA and protein? What is the mechanism of synchronization of crab clockwork and, as experiments discussed in Chapter 3 suggested might be the case, are there separate cellular or molecular clock-setting mechanisms for the various tidal time-cues? Even when questions such as these are answered there will inevitably be even more questions to address before living clockwork of tidal, semilunar, lunar and annual periodicities is finally characterized. Scientific research has the habit of raising more questions than it answers and, in its pursuance the search for living clockwork in marine organisms has clearly changed its focus from behaviour to physiology and on to molecular biology. Residual periodic variables in the environment, once perceived as crucial time cues in Chronobiology, now seem even less relevant than ever. Lamont Cole's (1957) unicorn may not be finally tamed but it is safely corralled, thanks ultimately to the use of the powerful techniques of modern molecular biology. Moreover, moon-related rhythmicity in animal behaviour has indeed moved out of the realms of mysticism into the world of hard science and testable hypotheses, as Hauenschild (1960) predicted in the mid-twentieth century after his pioneering studies of the rhythmic nuptial dances of marine worms.

References

Abello, P., Reid, D. G. and Naylor, E. (1991). Comparative locomotor activity patterns in the portunid crabs *Liocarcinus holsatus* and *L. depurator*. *Journal of the Marine Biological Association UK*, **71**, 1–10.

Abello, P., Warman, C. G. and Naylor, E. (1997). Circatidal moulting rhythms in the shore crab *Carcinus maenas*. *Journal of the Marine Biological Association UK*, **77**, 277–280.

Aguzzi, J. and Sarda, F. (2008). A history of recent advancements of *Nephrops norvegicus* behavioural and physiological rhythms. *Review of Fish Biology and Fisheries*, **18**, 235–248.

Aguzzi, J., Sarda, F., Abello, P., Company, J. B. and Rotllant, G. (2003). Diel and seasonal patterns of *Nephrops norvegicus* (Decapoda: Nephropidae) catchability in the western Mediterranean. *Marine Ecology Progress Series*, **258**, 201–211.

Aguzzi, J., Company, J. B. and Garcia, J. A. (2008). The circadian behavioural regulation of the shrimp *Processa canaliculata* Leach 1851 (Decapoda, Processidae) in relation to depth, ontogeny and the reproductive cycle. *Crustaceana*, **81**, 1301–1316.

Aguzzi, J., Sanchez-Pardo, J., Garcia, J. A. and Sarda, F. (2009). Day-night and depth differences in haemolymph melatonin of the Norway Lobster, *Nephrops norvegicus* (L.). *Deep Sea Research*, doi:10.1016/j.dsr.2009.06.001.

Al-Adhub, A. H. Y. and Naylor, E. (1977). Daily variations in *Dichelopandalus bonnieri* (Caullery) as a component of the epibenthos. *European Marine Biology Symposia*, **11**, 1–6.

Aldrich, J. C. (1997). Crab clocks sent for calibration. *Chronobiology International*, **14**, 435–437.

Alheit, J. and Naylor, E. (1976). The behavioural basis of intertidal zonation in *Eurydice pulchra* Leach. *Journal of Experimental Marine Biology and Ecology*, **23**, 135–144.

Ameyaw-Akumfi, C. and Naylor, E. (1987a). Temporal patterns of shell-gape in *Mytilus edulis*. *Marine Biology*, **95**, 237–242.

Ameyaw-Akumfi, C. and Naylor, E. (1987b). Spontaneous and induced components of salinity preference behaviour in *Carcinus maenas*. *Marine Ecology Progress Series*, **37**, 153–158.

Andersson, S., Kautsky, L. and Kalvas, A. (1994). Circadian and lunar gamete release in *Fucus vesiculosus* in the atidal Baltic Sea. *Marine Ecology Progress Series*, **110**, 195–201.

Arechiga, H. (1977). Circadian rhythmicity in the nervous system of crustaceans. *Federation Proceedings*, **36(7)**, 2036–2041.

Arechiga, H. (1993). Circadian systems. *Current Opinion in Neurobiology*, **3**, 1005–1010.

Arechiga, H. and Huberman, A. (1980). Hormonal modulation of neuronal activity in crustaceans. In *Comparative Aspects of Neuro-endocrine Control of Behaviour*, eds. C. Valverde and H. Arechiga. Basel: S. Karger, pp. 16–34.

Arechiga, H. and Wiersma, C. A. G. (1969). Circadian rhythm of responsiveness in crayfish visual units. *Journal of Neurobiology*, **1**, 71–85.

Arechiga, H., Huberman, A. and Naylor, E. (1974). Hormonal modulation of circadian neural activity in *Carcinus maenas* (L.). *Proceedings of the Royal Society of London, B*, **187**, 299–313.

Arechiga, H., Huberman, A. and Martinez-Palomo, A. (1977). Release of a neuro-depressing hormone from the crustacean sinus gland. *Brain Research*, **128**, 93–108.

Arechiga, H., Williams, J. A., Pullin, R. S. V. and Naylor, E. (1979). Cross-sensitivity to neuro-depressing hormone and its effect on locomotor rhythmicity in two different groups of crustaceans. *General and Comparative Endocrinology*, **37**, 350–357.

Arechiga, H., Atkinson, R. J. A. and Williams, J. A. (1980). Neurohumoral basis of circadian rhythmicity in *Nephrops norvegicus* (L.). *Marine Behaviour and Physiology*, **7**, 185–197.

Arechiga, H., Garcia, U. and Rodriguez-Sosa, L. (1985). Neurosecretory role of crustacean eyestalk in the control of neuronal activity. In *Model Neural Networks and Behaviour*, ed. A. I. Selverston. New York: Plenum Publishing Co., pp. 361–379.

Arendt, J., Deacon, S., English, J., Hampton, S. and Morgan, L. (1995). Melatonin and adjustment to phase-shift. *Journal of Sleep Research*, **4**, 74–79.

Arnoult, F., Vivien-Roels, B. and Vernet, G. (1994). Melatonin in the nermertine worm *Lineus lacteus*: identification and daily variations. *Biological Signals*, **3**, 296–301.

Arudpragasam, K. D. and Naylor, E. (1964). Gill ventilation volumes, oxygen consumption and respiratory rhythms in *Carcinus maenas* (L). *Journal of Experimental Biology*, **41**, 309–321.

Aschoff, J. (1960). Exogenous and endogenous components in circadian rhythms. *Cold Spring Harbor Symposia for Quantitative Biology*, **25**, 11–28.

Aschoff, J., von St-Paul, U. and Wever, R. (1971). The longevity of flies under the influence of time displacement. *Naturwissenschaften*, **58**, 574.

Atkinson, R. J. A. and Naylor, E. (1976). An endogenous activity rhythm and rhythmicity of catches of *Nephrops norvegicus* (L.). *Journal of Experimental Marine Biology and Ecology*, **25**, 95–108.

Ball, E. E. (1968). Activity patterns and retinal pigment migration in *Pagurus* (Decapoda: Paguridae). *Crustaceana*, **14**, 302–306.

Barbieri, C. and Rampazzi, F., eds. (2001). *Earth, Moon and Planets*, **85–86**, 575 pp.

Barham, E. G. (1963). Siphonophores and the deep scattering layer. *Science*, **140**, 826–828.

Barr, L. (1970). Diel vertical migration of *Pandalus borealis* in Kachemak Bay, Alaska. *Journal of the Fisheries Research Board of Canada*, **27**, 669–676.

Beentjes, M. P. and Williams, B. G. (1986). Endogenous circatidal rhythmicity in the New Zealand cockle *Chione stuchburyi* (Bivalvia: Veneridae). *Marine Behaviour and Physiology*, **12**, 171–180.

Benhamou, S., Bonadonna, F. and Jouventin, P. (2003). Successful homing of magnet-carrying white-chinned petrels released in the open sea. *Animal Behaviour*, **65**, 729–734.

Benn, C. R. (2001). The moon and the origin of life. *Earth, Moon and Planets*, **85–86**, 61–66.

Bennett, D. B. and Brown C. G. (1983). Crab (*Cancer pagurus*) migrations in the English Channel. *Journal of the Marine Biological Association UK*, **63**, 371–398.

Bentley, M. G., Olive, P. J. W. and Last, K. (2001). Sexual satellites, moonlight and nuptial dances of worms: the influence of the moon on reproduction of marine animals. *Earth, Moon and Planets*, **85–86**, 67–84.

Bergin, M. E. (1981). Hatching rhythms in *Uca pugilator*. *Marine Biology*, **63**, 151–158.

Berry, A. J. (1986). Semilunar and lunar spawning periodicity in some tropical littorinid gastropods. *Journal of Molluscan Studies*, **52**, 144–149.

Berrill, I. K., Porter, M. J. R., Smart, A., Mitchell, D. and Bromage, N. R. (2003). Photoperiodic effects on precocious maturation, growth and smoltification in Atlantic salmon, *Salmo salar. Aquaculture*, **222**, 239–252.

Bliss, D. E. (1982 onwards). *The Biology of Crustacea*, **1–10**. London: Academic Press.

Block, G. D. and Roberts, M. H. (1981). Circadian pacemaker in the *Bursatella* eye: properties of the rhythm and its effect on locomotor behaviour. *Journal of Comparative Physiology*, **142**, 403–410.

Boden, B. P. and Kampa, E. M. (1967). The influence of natural light on the vertical migration of an animal community in the sea. *Symposia of the Zoological Society of London*, **19**, 15–26.

Bogorov, V. G. (1946). Diurnal vertical migration of zooplankton in polar seas. *Trudy Institut Okeanologie An SSSR*, **1**, 151–158.

Bohn, G. (1903). Sur les movements oscillatoires des *Convoluta roscoffensis*. *Comptes rendues Académie des Sciences Paris*, **137**, 576–578.

Bohn, G. and Pieron H. (1906). Le rhythme des marees et la phenomene de l'anticipation reflexe. *Comptes rendues Societe de Biologie, Paris*, **61**, 660–661.

Bonadonna, F., Chamaillé-Jammes, S., Pinaud, D. and Weimerskirch, H. (2003a). Magnetic cues: Are they important in black–browed albatross *Diomedea melanophris* orientation? *Ibis*, **145**, 152–155.

Bonadonna, F., Benhamou, S. and Jouventin, P. (2003b). Orientation in 'featureless' environments: the extreme case of pelagic birds. In *Avian Migration*, eds. P. Berthold, E. Gwinner and E. Sonnenschein. Heidelberg: Springer.

Bonadonna, F., Bajzak, C., Benhamou, S., *et al.* (2005). Orientation in the wandering albatross: interfering with magnetic perception does not affect orientation performance. *Proceedings of the Royal Society of London, B*, **272**, 489–495.

Brady, J. (1979). *Biological Clocks*. London: Arnold, 60 pp.

Brady, J. (1982). Circadian rhythms in animal physiology. In *Biological Timekeeping*, ed. J. Brady. *Society for Experimental Biology Seminar Series*, **14**, 121–140.

Brady, J. (1987). Circadian rhythms: endogenous or exogenous? *Journal of Comparative Physiology A*, **161**, 711–714.

Brandstatter, R. (2002). The circadian pacemaking system of birds. In *Biological Rhythms*, ed. V. Kumar. New Delhi: Narose Publishing House.

Bregazzi, P. K. and Naylor, E. (1972). The locomotor activity rhythm of *Talitrus saltator* (Montagu). *Journal of Experimental Biology*, **57**, 375–391.

Broda, H. Brugg, D. Homma, K. L. and Hastings, J. W. (1985). Circadian communication between unicells? *Cell Biophysics*, **8**, 47–67.

Brown, F. A. (1954). Persistent activity rhythms in the oyster. *American Journal of Physiology*, **178**, 510–514.

Brown, F. A. (1958). The rhythmic nature of animals and plants. *Northwestern University Tri-Quarterly*, **1**, 35–47.

Brown, F. A. (1960). Response to pervasive geophysical factors and the biological clock problem. *Cold Spring Harbor Symposia on Quantitative Biology*, **25**, 57–71.

Brown, F. A. (1962a). *Biological Clocks*. American Institute for Biological Sciences, B.S.C.S. Pamphlet **2**. Boston, USA: Heath and Co, 36 pp.

Brown, F. A. (1962b). Extrinsic rhythmicality: a reference frame for biological rhythms under so-called constant conditions. *Annals of the New York Academy of Sciences*, **98(4)**, 775–787.

Brown, F. A. (1965). A unified theory for biological rhythms. In *Circadian Clocks: Proceedings of the Feldafing Summer School*, ed. J. Ascoff. Amsterdam: North Holland Publishing Co., pp. 231–261.

Brown, F. A., Fingerman, M., Sandeen, M. I. and Webb, H. M. (1953). Persistent diurnal and tidal rhythms of colour change in the fiddler crab *Uca pugnax*. *Journal of Experimental Zoology*, **123**, 29–60.

Brown, F. A., Webb, H. M., Bennett, M. F. and Sandeen, M. I. (1954). Temperature-independence of the frequency of the endogenous rhythm of *Uca*. *Physiological Zoology*, **17**, 346–349.

Brown, F. A., Freeland, R. O. and Ralph, C. L. (1955a). Persistent rhythms of oxygen consumption in potatoes, carrots and the seaweed, *Fucus*. *Plant Physiology*, **30(3)**, 280–292.

Brown, F. A., Webb, H. M. and Bennett, M. F. (1955b). Proof for an endogenous component in persistent solar and lunar rhythmicity in organisms. *Proceedings of the National Academy of Sciences, Washington*, **41**, 93–100.

Bryson, B. (2003). *A Short History of Nearly Everything*. Doubleday, 500 pp.

Bűnning, E. (1960). Circadian rhythms and time measurement in photoperiodism. *Cold Spring Harbor Symposia on Quantitative Biology*, **25**, 249–256.

Bűnning, E. (1967). *The Physiological Clock*, New York: Springer–Verlag, 167 pp.

Cahill, G. M. (2002). Clock mechanisms in zebra fish. *Cell and Tissue Research*, **309**, 27–34.

Carlisle, D. B. and Knowles, F. G. W. (1959). *Endocrine Control in Crustaceans*. Cambridge, UK: Cambridge University Press, 120 pp.

Carr, A. (1963). Orientation problems in the high seas travel and terrestrial movements of marine turtles. In *Bio-telemetry*, ed. L. E. Slater. New York: Pergamon Press, pp. 179–193.

Caspers, H. (1984). Spawning periodicity and habitat of the Palolo worm *Eunice viridis* in the Samoan Islands. *Marine Biology*, **79**, 229–236.

Castel, J. and Viega, J. (1990). Distribution and retention of the copepod *Eurytemora affinis hirundoides* in a turbid estuary. *Marine Biology*, **107**, 119–128.

Chandrashekaran, M. K. (1965). Persistent tidal and diurnal rhythms of locomotor activity and oxygen consumption in *Emerita asiatica*. *Zeitschrift fur Physiologie*, **50**, 137–150.

Chapin, J. P. and Wing, L. W. (1959). The Wideawake Calendar 1941–1958. *Auk*, **76**, 153–158.

Chapman, C. J. and Rice, A. L. (1971). Some direct observations on the ecology and behaviour of the Norway lobster, *Nephrops norvegicus*. *Marine Biology*, **10**, 321–329.

Christianson, R. and Sweeney, B. M. (1972). Sensitivity to stimulation, a component of the circadian rhythm in luminescence in *Gonyaulax*. *Plant Physiology*, **49**, 994–997.

Christy, J. H. (1978). Adaptive significance of reproductive cycles in the fiddler crab *Uca pugilator*: a hypothesis. *Science*, **199**, 453–455.

Chung, J. S. and Webster, S. G. (2006). Dynamics of *in vivo* release of moult-inhibiting hormone and crustacean hyperglycaemic hormone in the shore crab *Carcinus maenas*. *Endocrinology*, **146**, 5545–5551.

Clark, F. N. (1925). The life history of *Leuresthes tenuis*, an atherine fish with tide-controlled spawning habits. *California Fish and Game Committee Bulletin*, **10**, 1–51.

Clark, L. B. and Hess, H. W. (1940). Swarming of the Atlantic Palolo worm *Leodice fucata*. *Carnegie Publications*, **524**, 21–27.

Clark, R. B. (1965). Endocrinology and reproductive biology of polychaetes. *Oceanography and Marine Biology Annual Reviews*, **3**, 211–255.

Cole, L. C. (1957). Biological clock in the unicorn. *Science*, **125**, 874.

Colin, P. L., Shapiro, D. Y. and Weiler, D. (1987). Aspects of the reproduction of two species of groupers *Epinephelus guttatus* and *E. striatus* in the West Indies. *Bulletin of Marine Science*, **40**, 220–230.

Colman, J. S. (1950). *The Sea and its Mysteries*. London: G. Bell and Sons, 285 pp.

Conway Morris, S. (2003). *Life's Solution*. Cambridge, UK: Cambridge University Press, 464 pp.

Criales, M. M., Wang, J., Browder, J. A. and Robblee, M. B. (2005). Tidal and seasonal effects on transport of pink shrimp postlarvae. *Marine Ecology Progress Series*, **286**, 231–238.

Cronin, T. W. (1982). Estuarine retention of larvae of the crab *Rhithropanopeus harrisii*. *Estuarine, Coastal and Shelf Science*, **15**, 207–220.

Cronin, T. W. and Forward, R. B. (1979). Tidal vertical migrations and endogenous rhythm in estuarine crab larvae. *Science*, **205**, 1020–1022.

Cushing, D. H. (1951). The vertical migration of planktonic Crustacea. *Biological Reviews*, **26(2)**, 158–192.

Cushing, D. H. (1969). The regularity of the spawning season in some fishes. *Journal du Conseil pour l'exploration de la mer*, **33**, 81–92.

Daan, S. (1982). Circadian rhythms in animals and plants. In *Biological Timekeeping*, ed J. Brady. *Society for Experimental Biology Seminar Series*, **14**, 11–32.

Daase, M., Eiane, K., Aksues, D. L. and Vogedes, D. (2008). Vertical distribution of *Calanus* spp. and *Metridia longa* at four Arctic localities. *Marine Biology Research*, **4**, 193–207.

Dalley, R. (1979). Effects of non-circadian light cycles on survival and development of *Palaemon elegans* reared in the laboratory. *European Marine Biology Symposia*, **13**, 157–163.

Dalley, R. (1980a). Effects of non-circadian light/dark cycles on the growth and moulting of *Palaemon elegans* reared in the laboratory. *Marine Biology*, **56**, 71–78.

Dalley, R. (1980b). The survival and development of the shrimp *Crangon crangon* (L.), reared under non-circadian light/dark cycles. *Journal of Experimental Marine Biology and Ecology*, **47**, 101–112.

Darwin, C. (1859). *On the Origin of Species by Natural Selection*. London: John Murray, 491 pp.

Darwin, C. and Darwin, F. (1880). *The Power of Movement in Plants*. London: John Murray.

Davenport, J. (1985a). Osmotic control in marine animals. *Symposia of the Society for Experimental Biology*, **39**, 207–244.

Davenport, J. (1985b). *Environmental Stress and Behavioural Adaptations*. London and Sydney: Croom Helm, 122 pp.

DeCoursey, P. J. (1983). Biological timing. In *The Biology of Crustacea*, **7**, ed. D. E. Bliss. New York: Academic Press, pp. 107–162.

Della Santina, P. D. and Naylor, E. (1993). Endogenous rhythms in the homing behaviour of the limpet *Patella vulgata* Linnaeus. *Journal of Molluscan Studies*, **59**, 87–91.

De Pauw, N. (1973). On the distribution of *Eurytemora affinis* (Poppe) (Copepoda) in the western Scheldte estuary. *Verh. Int. Verein. theor. angew. Limnol.*, **18**, 1462–1472.

De Wilde, P. A. W. J. and Berghuis, E. M. (1979). Cyclic temperature fluctuations in a tidal mud flat. *European Marine Biology Symposia*, **13**, 435–441.

Dieck, I. T. (1991). Circannual growth rhythm and photoperiodic sorus induction in the kelp *Laminaria setchellii* (Phaeophyta). *Journal of Phycology*, **27**, 341–350.

Digby, P. S. B. (1961). The vertical distribution and movements of plankton under midnight sun conditions in Spitzbergen. *Journal of Animal Ecology*, **30(1)**, 9–25.

Dirckson, H., Webster, S. G. and Keller, R. (1988). Immunocytochemical demonstration of the neurosecretory systems containing putative moult-inhibiting hormone and hyperglycaemic hormone in the eyestalk of brachyuran crustaceans. *Cell and Tissue Research*, **251**, 3–12.

Dronkers, J. J. (1964). *Tidal Computations in Rivers and Coastal Waters*. New York: John Wiley and Sons.

Dunlap, J. C., Loros, J. J. and DeCoursey, P. (2004). *Chronobiology: Biological Timekeeping*. Sunderland, MA, USA: Sinauer Associates Inc. Publishing, 382 pp.

Dustan, J. and Bromage, N. (1991). Circannual rhythms of gonadal maturation in female rainbow trout (*Oncorhynchus mykiss*). *Journal of Biological Rhythms*, **6(1)**, 49–53.

Eaton, J. W. and Simpson, P. (1979). Vertical migrations of the intertidal dinoflagellate *Amphidinium herdmaniae* Kofoid and Swezy. *European Marine Biology Symposia*, **13**, 339–345.

Eckert, R. (1966). Excitation and luminescence in *Noctiluca miliaris*. In *Bioluminescence in Progress*, eds. F. H. Johnson and Y. Haneda. Princeton, NJ, USA: Princeton University Press, pp. 269–300.

Edwards, J. M. and Naylor, E. (1987). Endogenous circadian changes in orientational behaviour in *Talitrus saltator*. *Journal of the Marine Biological Association UK*, **67**, 17–26.

Edwards, R. R. C. (1978). The fishery and fishery biology of penaeid shrimp on the Pacific coast of Mexico. *Oceanography and Marine Biology Annual Reviews*, **16**, 145–180.

Ehrenbaum, E. (1911). Mittelungen uber die Lebensverhaltnisse unserer Fische. 9: *Pleuronectes flesus* L., Flunder, Butt. *Der Fischerbote*, **4(3)**, 109–176.

Ehret, C. F. and Trucco, E. (1967). Molecular models of the circadian clock: 1. The chronon concept. *Journal of Theoretical Biology*, **15**, 240–262.

Emata, A. C., Meier, A. H. and Hsiao, S. M. (1991). Daily variations in plasma hormone concentrations during the semilunar spawning cycle of the Gulf killifish, *Fundulus grandis*. *Journal of Experimental Zoology*, **259**, 343–354.

Ennis, G. P. (1973). Endogenous rhythmicity associated with larval hatching in the lobster *Homarus gammarus*. *Journal of the Marine Biological Association UK*, **53**, 531–538.

Enright, J. T. (1963). The tidal activity rhythm of a sand-beach amphipod. *Zeitschrift fur vergleichende Physiologie*, **46**, 276–313.

Enright, J. T. (1965a). The search for rhythmicity in biological time-series. *Journal of Theoretical Biology*, **8**, 426–468.

Enright, J. T. (1965b). Entrainment of a tidal rhythm. *Science, NY*, **147**, 864–867.

Enright, J. T. (1971). The internal clock of drunken isopods. *Zeitschrift fur vergleichende Physiologie*, **75**, 332–346.

Enright, J. T. (1972). A virtuoso isopod: circalunar rhythms and their fine structure. *Journal of Comparative Physiology*, **77**, 141–162.

Enright, J. T. (1974). Rhythmicity. (Review of *The Control of Physiological and Behavioral Tidal Rhythms*, by John D. Palmer, 1974), *Science*, **185**, p. 521.

Enright, J. T. (1975). Orientation in time: endogenous clocks. In *Marine Ecology 2(2): Physiological Mechanisms*, ed. O. Kinne. Wiley Interscience, pp. 917–944.

Enright, J. T. (1976a). Resetting a tidal clock: a phase-response curve for *Excirolana*. In *Biological Rhythms in the Marine Environment*, ed. P. J. DeCoursey. SC, USA: University of South Carolina, pp. 103–114.

Enright, J. T. (1976b). Plasticity in an isopod's clockwork: shaking shapes form and affects phase and frequency. *Journal of Comparative Physiology*, **107**, 13–37.

Enright, J. T. (1977). Diel vertical migration: adaptive significance and timing. Part 1. Selective advantage: a metabolic model. *Limnology and Oceanography*, **22**, 856–872.

Enright, J. T. and Hamner, W. M. (1967). Vertical migration and endogenous rhythmicity. *Science*, **157**, 937–941.

Epifanio, C. E., Valenti, C. C. and Pembroke, A. E. (1984). Dispersal and recruitment of blue crab larvae in Delaware Bay, USA. *Estuarine, Coastal and Shelf Science*, **18**, 1–12.

Esterly, C. O. (1911). Diurnal migration of *Calanus finmarchicus* in the San Diego region during 1909. *International Review of Hydrobiology*, **4**, 140–151.

Esterly, C. O. (1917). The occurrence of a rhythm in the geotropism of two species of plankton copepods when certain recurring external conditions are absent. *University of California Publications in Zoology*, **16**, 393–400.

Everson, I. (1982). Diurnal variations in mean volume back-scattering strength of an Antarctic krill (*Euphausia superba*) patch. *Journal of Plankton Research*, **4**, 155–162.

Farmer, A. S. D. (1974). Field assessments of diurnal activity in Irish Sea populations of the Norway lobster *Nephrops norvegicus* (L.) (Decapoda: Nephropidae). *Estuarine and Coastal Marine Science*, **2**, 37–47.

Fingerman, M. (1955). Persistent daily and tidal rhythms of color change in *Callinectes sapidus*. *Biological Bulletin*, **109(2)**, 255–264.

Follett, B. K. (1982). Photoperiodic physiology in animals. In *Biological Timekeeping*, ed. J. Brady, *Society for Experimental Biology Seminar Series*, **14**, 83–100.

Forward, R. B. (1980). Phototaxis of a sand-beach amphipod: physiology and tidal rhythms. *Journal of Comparative Physiology*, **135**, 243–250.

Forward, R. B. (1988). Diel vertical migration: zooplankton photobiology and behaviour. *Oceanography and Marine Biology Annual Reviews*, **26**, 361–393.

Forward, R. B. (2009). Larval biology of the crab *Rhithropanopeus harrisii* (Gould): a synthesis. *Biological Bulletin* 216(3), 243–256.

Forward, R. B. and Cronin, T. W. (1980). Tidal rhythms of activity and phototaxis of an estuarine crab larva. *Biological Bulletin*, **158**, 295–303.

Forward, R. B. and Tankersley, R. A. (2001). Selective tidal-stream transport of marine animals. *Oceanography and Marine Biology Annual Reviews*, **39**, 305–353.

Forward, R. B., Cronin, T. W. and Stearns, D. E. (1984). Control of diel vertical migration photoresponses of a larval crustacean. *Limnology and Oceanography*, **29**, 146–154.

Forward, R. B., Cohen, J. H., Irvine, R. D. *et al.* (2004). Settlement of blue crab *Callinectes sapidus* megalopae in a North Carolina estuary. *Marine Ecology Progress Series*, **269**, 237–247.

Foster, R. and Kreitzman, L. (2005). *Rhythms of Life*. London: Profile Books, 278 pp.

Frasseto, R., Backus, R. H. and Hays, E. (1962). Sound scattering layers and their relation to thermal structures in the Strait of Gibraltar. *Deep Sea Research*, **9(1)**, 69–72.

Frazer, F. C. (1937). On the development and distribution of the young stages of krill (*Euphausia superba*). *Discovery Reports*, **14**, 1–192.

Gambinieri, S. and Scapini, F. (2008). Importance of orientation to the sun and local landscape features in young inexpert *Talitrus saltator* (Amphipoda: Talitridae) from two Italian beaches differing in morphodynamics, erosion and stability. *Estuarine, Coastal and Shelf Science*, **77**, 357–368.

Gamble, F. W. and Keeble, F. (1903). The bionomics of *Convoluta roscoffensis*, with special reference to its green cells. *Proceedings of the Royal Society of London, B*, **72**, 93–98.

Gibson, R. N. (1965). Rhythmic activity in littoral fish. *Nature, London*, **207**, 544–545.

Gibson, R. N. (1967). Experiments on the tidal rhythm of *Blennius pholis*. *Journal of the Marine Biological Association, UK*, **47**, 97–111.

Gibson, R. N. (1969). Activity rhythms in two species of *Blennius* from the Mediterranean. *Vie et Milieu*, **XX** 1A, 235–244.

Gibson, R. N. (1973). Tidal and circadian activity rhythms in juvenile plaice *Pleuronectes platessa*. *Marine Biology*, **22**, 379–386.

Gibson, R. N. (2004). Go with the flow: tidal migrations in marine animals. *Hydrobiologia*, **503**, 153–161.

Gibson, R. N., Blaxter, J. H. S. and De Groot, S. J. (1978). Developmental changes in the activity rhythm of plaice (*Pleuronectes platessa*). In *Rhythmic Activity of Fishes*, ed. J. E. Thorpe. London: Academic Press, pp. 169–186.

Grubb, T. C. (1974). Olfactory navigation to nesting burrows in Leach's petrel. *Auk*, **90**, 78–82.

Gwinner, E. (1989). Photoperiod as a modifying and limiting factor in the expression of avian circannual rhythms. *Journal of Biological Rhythms*, **4**, 237–256.

Hall, J. C. (2003). Genetics and molecular biology of rhythms in *Drosophila* and other insects. *Advances in Genetics*, **48**, 1–280.

Hamner, W. M. (1988). Behavior of plankton and patch formation in pelagic ecosystems. *Bulletin of Marine Science*, **43**, 752–757.

Hardeland, R. and Poeggeler, B. (2003). Non-vertebrate melatonin. *Journal of Pineal Research*, **34**, 233–241.

Hardeland, R., Balzer, I., Poeggeler, B. *et al.* (1995). On the primary functions of melatonin in evolution: modulation of photoperiodic signals in a unicell, photooxidation and scavenging of free radicals. *Journal of Pineal Research*, **18**, 104–111.

Hardin, P. E., Hall, J. C. and Rosbash, M. (1990). Feedback of the *Drosophila* period gene product on the circadian cycling of its messenger RNA levels. *Nature*, **343**, 536–540.

Hardy, A. C. (1956). *The Open Sea: World of Plankton*. London: Collins, 335 pp.

Hardy, A. C. and Gunther, E. R. (1935). The plankton of the South Georgia whaling ground and adjacent waters, 1926–7. *Discovery Reports*, **11**, 511–538.

Hardy, A. C. and Paton, W. N. (1947). Experiments on the vertical migration of plankton animals. *Journal of the Marine Biological Association UK*, **26**, 467–526.

Harker, J. E. (1964). The physiology of diurnal rhythms. London: Cambridge University Press, 114 pp.

Harris, G. J. and Morgan, E. (1984a). The effects of salinity changes on the endogenous circatidal rhythm of the amphipod *Corophium volutator* (Pallas). *Marine Behaviour and Physiology*, **10**, 199–217.

Harris, G. J. and Morgan, E. (1984b). The effects of ethanol, valinomycin and cyclohexamide on the endogenous circatidal rhythm of the estuarine amphipod *Corophium volutator* (Pallas). *Marine Behaviour and Physiology*, **10**, 219–233.

Harris, G. J. and Morgan, E. (1984c). The location of circatidal pacemakers in the estuarine amphipod crustacean *Corophium volutator* using a selective chilling technique. *Journal of Experimental Biology*, **110**, 125–142.

Harris, J. E. (1963). The role of endogenous rhythms in vertical migration. *Journal of the Marine Biological Association UK*, **43**, 153–166.

Hastings, J. W. and Sweeney, B. M. (1957). On the mechanism of temperature-independence in a biological clock. *Proceedings of the National Academy of Science USA*, **48**, 804–811.

Hastings, J. W. and Sweeney, B. M. (1958). A persistent diurnal rhythm of luminescence in *Gonyaulax polyedra*. *Biological Bulletin*, **115**, 440.

Hastings, J. W., Astrachan, L. and Sweeney, B. M. (1961). A persistent daily rhythm of photosynthesis. *Journal of General Physiology*, **45**, 69–76.

Hastings, M. H. (1981a). The entraining effect of turbulence on the circatidal activity rhythm and its semilunar modulation in *Eurydice pulchra*. *Journal of the Marine Biological Association UK*, **61**, 151–160.

Hastings, M. H. (1981b). Semilunar variations of endogenous circatidal rhythms of activity and respiration in the isopod *Eurydice pulchra*. *Marine Ecology Progress Series*, **4**, 85–90.

Hauenschild, C. (1960). Lunar periodicity. *Cold Spring Harbor Symposia on Quantitative Biology*, **25**, 491–497.

Hawking, S. (1988). *A Brief History of Time*. London: Bantam Press, 218 pp.

Hays, G. C. (1995). Zooplankton avoidance activity. *Nature*, **376**, 650.

Hays, G. C., Åkesson, S., Broderick, A. C. *et al.* (2003). Island-finding ability of marine turtles. *Proceedings of the Royal Society of London B*, **270**, Supp. 1, 5–7.

Head, J. W. (2001). Lunar and planetary perspectives on the geological history of the earth. *Earth Moon and Planets*, **85–86**, 153–177.

Heckman, C. W. (1986). The anadromous migration of a calanoid copepod, *Eurytemora affinis* (Poppe, 1880) in the Elbe estuary. *Crustaceana*, **50**, 176–181.

Herman, S. S. (1963). Vertical migration of the opossum shrimp, *Neomysis americana*. *Limnology and Oceanography*, **8**, 228–238.

Herring, P. J. and Roe, H. S. J. (1988). The photoecology of pelagic decapods. *Symposia of the Zoological Society of London*, **59**, 263–290.

Hersey, J. B. and Moore, H. B. (1948). Progress report on scattering layer observations in the Atlantic Ocean. *Transactions of the American Geophysical Union*, **29(3)**, 341–354.

Hill, A. E. (1991a). A mechanism for horizontal zooplankton transport by vertical migration in tidal currents. *Marine Biology*, **111**, 485–492.

Hill, A. E. (1991b). Vertical migration in tidal currents. *Marine Ecology Progress Series*, **75**, 39–54.

Hill, A. E. (1994). Horizontal zooplankton dispersal by diel vertical migration in solar tidal currents on the northwest European continental shelf. *Continental Shelf Research*, **14(5)**, 491–506.

Hoffman, K. (1982). Time-compensated celestial orientation. In *Biological Timekeeping*, ed. J. Brady. *Society for Experimental Biology Seminar Series*, **4**, 49–62.

Hopkins, C. C. E. and Gulliksen, B. (1978). Diurnal vertical migration and zooplankton-epibenthos relationships in a north Norwegian fjord. *European Marine Biology Symposia*, **12**, 271–280.

Hough, A. R. and Naylor, E. (1991). Field studies on retention of the planktonic copepod *Eurytemora affinis* in a mixed estuary. *Marine Ecology Progress Series*, **76**, 115–122.

Hough, A. R. and Naylor, E. (1992). Endogenous rhythms of circatidal swimming activity in the estuarine copepod *Eurytemora affinis* (Poppe). *Journal of Experimental Biology and Ecology*, **161**, 27–32.

Howie, D. I. D. (1984). The reproductive biology of the lugworm *Arenicola marina* L. *Fortschrifte der Zoologie*, **29**, 247–263.

Hsiao, S. M. and Meier, A. H. (1989). Comparison of semilunar cycles of spawning activity in *Fundulus grandis* and *F. heteroclitus* held under constant laboratory conditions. *Journal of Experimental Zoology*, **252**, 213–218.

Hsiao, S. M. and Meier, A. H. (1992). Free-running circasemilunar spawning rhythm of *Fundulus grandis* and its temperature compensation. *Fish Physiology and Biochemistry*, **10(3)**, 259–265.

Hsiao, S. M., Greeley, M. S. J. and Wallace, R. A. (1994). Reproductive cycling in female *Fundulus heteroclitus*. *Biological Bulletin*, **186**, 271–284.

Huberman, A. (1996). Neurodepressing hormone (NDH): fact or fiction? *Crustaceana* **69**, 1–18.

Huberman, A., Arechiga, H., Cimet, A. de la Rosa, J. and Aramburo, C. (1979). Isolation and purification of a neurodepressing hormone from the eyestalk of *Procambarus bouvieri* Ortmann. *European Journal of Biochemistry*, **99**, 203–208.

Hunter, E., Metcalfe, J. D., O'Brian, C. M., Arnold, G. P. and Reynolds, J. D. (2004). Vertical activity patterns of free-swimming adult plaice in the southern North Sea. *Marine Ecology Progress Series*, **279**, 261–273.

Hughes, R. N. (1983). Evolutionary ecology of reef organisms, with particular reference to corals. *Biological Journal of the Linnaean Society*, **20**, 39–58.

Irigoien, X., Conway, D. V. P. and Harris, R. P. (2004). Flexible diel vertical migration behaviour of zooplankton in the Irish Sea. *Marine Ecology Progress Series*, **267**, 85–97.

Itoh, M. T. and Sumi, Y. (1998). Melatonin and serotonin N-acetyltransferase activity in developing eggs of the cricket *Gryllus bimaculatus*. *Brain Research*, **81**, 90–99.

Iwasaki, H., Williams, S. B., Kitayama, Y. *et al.* (2000). A *kai C*–interacting histidine kinase, *sosa*, necessary to sustain robust circadian oscillation in cyanobacteria. *Cell*, **101**, 223–233.

Jacklett, J. W. (1969). Circadian rhythm of optic nerve impulses recorded in darkness from the isolated eye of *Aplysia*. *Science*, **164**, 562–563.

Jacklett, J. W. (1974). The effects of constant light and light pulses on the circadian rhythm in the eye of *Aplysia*. *Journal of Comparative Physiology*, **90**, 33–45.

Jacklett, J. W. (1982). Circadian clock mechanisms. In *Biological Timekeeping*, ed J. Brady. *Society for Experimental Biology Seminar Series*, **14**, 173–188.

Jager, Z. (1998). Accumulation of flounder larvae *Platichthys flesus* (L.) in the Dollard (Ems estuary), Wadden Sea. *Journal of Sea Research*, **40**, 43–57.

Jager, Z. (1999a). *Floundering: processes of tidal transport and accumulation of larval flounder* (Platichthys flesus *L.*) *in the Ems–Dollard nursery*. Wageningen: Ponsen and Looijen, 192 pp.

Jager, Z. (1999b). Selective tidal stream transport of flounder larvae (*Platichthys flesus* L.) in the Dollard (Ems estuary). *Estuarine Coastal and Shelf Science*, **49**, 347–362.

Jeffs, A., Tolimieri, N. and Montgomery, J. C. (2003). Crabs on cue for the coast: the use of underwater sound for orientation by pelagic crabs. *Marine and Freshwater Research*, **54**, 841–845.

Johannes, R. E. (1978). Reproductive strategies of coastal marine fishes in the tropics. *Environmental Biology of Fishes*, **3**, 65–84.

Johnson, C. H., Roeber, J. and Hastings, J. W. (1984). Circadian changes in enzyme concentration account for rhythm of enzyme activity in *Gonyaulax*. *Science*, **223**, 1428–1430.

Johnson, C. R. (2001). Endogenous timekeepers in photosynthetic organisms. *Annual Review of Physiology*, **63**, 695–728.

Jones, S. (1999). *Almost Like a Whale*. London: Doubleday, 499 pp.

Jones, D. A. and Hobbins, C. St. C. (1985). The role of biological rhythms in some sand-beach isopod cirolanid crustaceans. *Journal of Experimental Marine Biology and Ecology*, **93**, 47–59.

Jones, D. A. and Naylor, E. (1970). The swimming rhythm of the sand-beach isopod *Eurydice pulchra*. *Journal of Experimental Biology and Ecology*, **4**, 188–199.

Kain, J. M. (1989). The seasons in the subtidal. *British Phycological Journal*, **24**, 203–215.

Kando, T., Tsinoreman, N. F., Golden, S. S. *et al.* (1994). Circadian clock mutants of cyanobacteria. *Science*, **266**, 1233–1236.

Karban, R. (1982). Increased reproductive success at high densities and predator satiation for periodical cicadas. *Ecology*, **63**, 321–328.

Keeble, F. (1910). *Plant Animals: A Study in Symbiosis.* Cambridge, UK: Cambridge University Press, 163 pp.

Kelly, P., Sulkin, S. D. and Heukelem, L. van. (1982). A dispersal model for larvae of the deep sea red crab *Geryon quinquidens* based upon behavioral regulation of vertical migration in the hatching stage. *Marine Biology*, **72**, 35–43.

Kennedy, B. and Pearse, J. S. (1975). Lunar synchronization of the monthly reproductive rhythm in the sea urchin *Centrostephanus coronatus* Verrill. *Journal of Experimental Marine Biology and Ecology*, **17**, 323–331.

Kennedy, F., Naylor, E. and Jaramillo, E. (2000). Ontogenetic differences in the circadian locomotor activity rhythm of the talitrid amphipod crustacean *Orchestoidea tuberculata. Marine Biology*, **137**, 511–517.

Klapow, L. A. (1976). Tidal and lunar rhythms of an intertidal crustacean. In *Biological Rhythms in the Marine Environment*, ed. P. J. DeCoursey. SC Columbia, USA: University of South Carolina Press, pp. 215–224.

Kleinhoonte, A. (1928). De door het licht geregelde autonome bewegwigen der *Canavalia bladeren.* Ph.D. Thesis, University of Delft.

Kluge, M. (1982). Biochemical rhythms in plants. In *Biological Timekeeping*, ed. J. Brady. *Society for Experimental Biology Seminar Series*, **14**, 159–172.

Knight-Jones, E. W. (1952). Reproduction of oysters in the Rivers Crouch and Roach, Essex, during 1947, 1948 and 1949. *Ministry of Agriculture, Fisheries and Food, Fisheries Investigations, Series II*, **18(2)**, 1–43.

Knight-Jones, E. W. and Morgan, E. (1966). Responses of marine animals to changes in hydrostatic pressure. *Oceanography and Marine Biology Annual Reviews*, **4**, 267–299.

Knights, A. M., Crow, T. P. and Burnell, G. (2006). Mechanisms of larval transport: vertical distribution of bivalve larvae varies with tidal conditions. *Marine Ecology Progress Series*, **326**, 167–174.

Konopka, R. J. and Benzer, S. (1971). Clock mutants in *Drosophila melanogaster. Proceedings of the National Academy of Sciences USA*, **68**, 2112–2116.

Korringa, P. (1957). Lunar periodicity. *Memoirs of the Geological Society of America*, **67**, 917–934.

Kramer, G. (1952). Experiments on bird migration. *Ibis*, **94**, 265–285.

Larimer. J. L. and Smith, J. T. F. (1980). Circadian rhythm of retinal sensitivity in crayfish: modulation by cerebral and optic ganglia. *Journal of Comparative Physiology*, **136**, 313–326.

Last, K. S., Bailhache, T., Kramer, C., Kiriacou, C. P. *et al.* (2009). Tidal, daily and lunar-day activity cycles in the marine polychaete *Nereis virens. Chronobiology International* **26**(2), 167–183.

Lee, J. W. and Williamson, D. I. (1975). The vertical distribution and diurnal migration of copepods at different stations in the Irish Sea. *Journal of the Oceanographical Society of Japan*, **31**, 199–211.

Lickey, M. E. and Wozniak, J. (1979). Circadian organization in *Aplysia* explored with red light, eye removal and behavioural recording. *Journal of Comparative Physiology*, **131**, 169–177.

Lockwood, A. P. M. (1976). Physiological adaptation to life in estuaries. In *Adaptation to Environment: Essays on the Physiology of Marine Animals*, ed. R. C. Newell. London: Butterworths, pp. 315–392.

Lohmann, K. J. and Lohmann, C. M. F. (2006). Sea turtles, lobsters and oceanic maps. *Marine and Freshwater Behaviour and Physiology*, **39**, 49–64.

Lohmann, K. J. and Willows, A. O. D. (1987). Lunar-modulated geomagnetic orientation by a marine mollusc. *Science*, **235**, 331–334.

Lohmann, K. J., Lohmann, C. F., Erhart, L. M., Bagley, D. A. and Swing, T. (2004). Geomagnetic map used in sea-turtle navigation. *Nature*, **428**, 909–910.

Longhurst, A. R. (1976). Vertical migration. In *The Ecology of the Seas*, eds. D. H. Cushing and J. J. Walsh. Oxford, UK: Blackwell, pp. 116–137.

Lüning, K. (1991). Circannual growth rhythm in a brown alga *Pterygophora californica*. *Botanica Acta*, **104**, 157–162.

Lüning, K. (2005). Endogenous rhythms and day-length effects on macroalgal development. In *Algal Culturing Techniques*, ed R. A. Anderson. London: Academic Press/Elsevier, pp. 347–364.

Lüning, K. and Kadel, P. (1993). Daylength range for circannual rhythmicity in *Pterygophora californica* (Alariaceae, Phaeophyta) and synchronization of seasonal growth by daylength cycles in several other brown algae. *Phycologia*, **32**, 379–387.

Lynch, B. R. and Rochette, R. (2007). Circatidal rhythm of free-roaming sub-tidal green crabs *Carcinus maenas*, revealed by radio-acoustic telemetry. *Crustaceana*, **80**, 345–355.

Macpherson, E. and Raventos, N. (2006). Relationship between pelagic larval duration and geographical distribution of Mediterranean littoral fishes. *Marine Ecology Progress Series*, **327**, 257–265.

Makarov, V. N., Schoschina, E. V. and Lüning, K. (1995). Diurnal and circadian periodicity of mitosis and growth in marine algae. I. Juvenile sporophytes of Laminariales (Phaeophyta). *European Journal of Phycology*, **30**, 261–266.

Marshall, S. M. and Orr, A. P. (1955). *The Biology of a Marine Copepod*. London: Oliver and Boyd, 195 pp.

Marta-Almeida, M., Dubert, J, Peliz, A. and Queiroga, H. (2006). Influence of vertical migration pattern on retention of crab larvae in a seasonal upwelling system. *Marine Ecology Progress Series*, **307**, 1–19.

Martin, H and Martin, U. (1987). Transfer of a time-signal isochronous with local time in translocation experiments to the geographical longitude. *Journal of Comparative Physiology* A, **160**, 3–9.

Martin, L. (1907). La memoire chez *Convoluta roscoffensis*. *Comptes rendu hebdomadaire des séances de l'Acadamie des sciences*, **145**, 555–557.

Massa, B., Benvenuti, P., Lo Valvo, M. and Papi, F. (1991). Homing of Cory's shearwaters (*Calonectris diomedea*) carrying magnets. *Bollettino di Zoologia*, **58**, 245–247.

Matthews, G. V. T. (1968). *Bird Migration*. Cambridge, UK: Cambridge University Press.

McLachlan, A., Wooldridge, T. and van der Horst, G. (1979). Tidal movements of the macrofauna on an exposed sandy beach in South Africa. *Journal of Zoology, London*, **188**, 433–442.

McGaw, I. J. and Naylor, E. (1992). Salinity preference behaviour of the shore crab *Carcinus maenas* in relation to colouration during intermoult and to prior acclimation. *Journal of Experimental Marine Biology and Ecology*, **155**, 145–159.

McLusky, D. M. (1981). *The Estuarine Ecosystem*. Glasgow and London: Blackie, 150 pp.

Menaker, M. (1976). A 'terrestrial' biologist looks at biorhythms in the marine environment. In *Biological Rhythms in the Marine Environment*, ed. P. J. DeCoursey. Columbia, SC, USA: University of South Carolina Press, pp. 273–278.

Menaker, M. and Tosini, G. (1996). The evolution of vertebrate circadian systems. In *Circadian Organization and Oscillatory Coupling. Proceedings of the VIth Sapporo*

Symposium on Biological Rhythms. Sapporo, Japan: Hokkaido University Press, pp. 39–52.

Meschini, E., Gagliardo, A. and Papi, F. (2008). Lunar orientation in sandhoppers is affected by shifting both the moon phase and the daily clock. *Animal Behaviour*, **76**, 25–35.

Metcalfe, J. D., Arnold, G. P. and Webb, P. W. (1990). The energetics of migration by selective tidal stream transport: an analysis of plaice tracked in the southern North Sea. *Journal of the Marine Biological Association UK*, **70**, 149–162.

Metcalfe, J. D., Hunter, E. and Buckley, A. A. (2006). The migratory behaviour of North Sea plaice: currents, clocks and clues. *Marine and Freshwater Behaviour and Physiology*, **39(1)**, 25–36.

Mezzetti, M. C., Naylor, E. and Scapini, F. (1994). Rhythmic responsiveness to visual stimuli in different populations of talitrid amphipods from Atlantic and Mediterranean coasts: an ecological interpretation. *Journal of Experimental Marine Biology and Ecology*, **181**, 279–291.

Michael, E. L. (1911). Classification and vertical distribution of the chaetognaths of the San Diego region. *University of California Publications in Zoology*, **8**, 21.

Mitsui, A., Kumazawa, S., Takahashi, A., Ikemoto, H. and Arai, T. (1986). Strategy by which nitrogen-fixing unicellular cyanobacteria grow photoautotrophically. *Nature*, **323**, 720–722.

Moore, H. B. and Corwen, E. G. (1956). The effects of temperature, illumination and pressure on the vertical distribution of zooplankton. *Bulletin of Marine Science of the Gulf and Caribbean*, **6**, 273–287.

Morgan, E. (1965). The activity rhythm of the amphipod *Corophium volutator* (Pallas) and its possible relationship to changes in hydrostatic pressure associated with the tides. *Journal of Animal Ecology*, **34**, 731–746.

Morgan, E. (2001). The moon and life on earth. *Earth, Moon and Planets*, **85–86**, 279–290.

Morgan, S. G. (1996). Influence of tidal variations on reproductive timing. *Journal of Experimental Marine Biology and Ecology*, **206**, 237–251.

Morgan, S. G. and Christy, J. H. (1994). Plasticity, constraint and optimality in reproductive timing. *Ecology*, **75**, 2185–2203.

Morgan, S. G. and Christy, J. H. (1995). Adaptive significance of the timing of larval release by crabs. *American Nature*, **145**, 457–479.

Morgan, S. G. and Christy, J. H. (1997). Planktivorous fishes as selective agents for reproductive synchrony. *Journal of Experimental Marine Biology and Ecology*, **209**, 89–101.

Morse, D., Fritz, L. and Hastings, J. W. (1990). What is the clock? Translational regulation of circadian bioluminescence. *Trends in Biochemical Science*, **15**, 262–265.

Müller, D. (1962). Uber jahres und lunarperiodische Erscheinungen bei einigen Braunalgen. *Botanica Marina*, **4**, 140–155.

Munro Fox, H. (1923). Lunar periodicity in reproduction. *Proceedings of the Royal Society of London, B*, **155**, 523–549.

Munro Fox, H. (1928). *SELENE, or Sex and the Moon*. London: Kegan Paul, Trench, Trubener and Co., 84 pp.

Murray, J. (1885). *Narrative Report of the Challenger Expedition*. London.

Naylor, E. (1958). Tidal and diurnal rhythms of locomotory activity in *Carcinus maenas* (L.). *Journal of Experimental Biology*, **35**, 602–610.

Naylor, E. (1960). Locomotory rhythms in *Carcinus maenas* (L.) from non-tidal conditions. *Journal of Experimental Biology*, **37**, 481–488.

Naylor, E. (1961). Spontaneous locomotor rhythm in Mediterranean *Carcinus*. *Pubblicazioni della Stazione Zoologica di Napoli*, **32**, 58–63.

Naylor, E. (1962). Seasonal changes in a population of *Carcinus maenas* (L.) in the littoral zone. *Journal of Animal Ecology*, **31**, 601–609.

Naylor, E. (1963). Temperature relationships of the locomotor rhythm of *Carcinus*. *Journal of Experimental Biology*, **40**, 669–679.

Naylor, E. (1976). Rhythmic behaviour and reproduction in marine animals. In *Adaptation to Environment*, ed. R. C. Newell. London: Butterworths, pp. 393–429.

Naylor, E. (1982). Tidal and lunar rhythms in animals and plants. In *Biological Timekeeping*, ed. J. Brady. *Society for Experimental Biology Seminar Series*, **14**, 33–48.

Naylor, E. (1985). Tidally rhythmic behaviour of marine animals. *Symposia of the Society for Experimental Biology*, **39**, 63–93.

Naylor, E. (1988). Rhythmic behaviour of decapod crustaceans. *Symposia of the Zoological Society of London*, **59**, 177–199.

Naylor, E. (1996). Crab clockwork: the case for interactive circatidal and circadian oscillators controlling rhythmic locomotor of *Carcinus maenas*. *Chronobiology International*, **13**, 153–161.

Naylor, E. (1997). Crab clocks re-wound. *Chronobiology International*, **14**, 427–430.

Naylor, E. (2001). Marine animal behaviour in relation to lunar phase. *Earth, Moon and Planets*, **85–86**, 291–302.

Naylor, E. (2002). Coastal animals that anticipate time and tide. *Ocean Challenge*, **11(3)**, 21–26.

Naylor, E. (2005). Chronobiology: implications for marine resource exploitation and management. *Scientia Marina*, **69**, 157–167.

Naylor, E. (2006). Orientation and navigation in coastal and estuarine zooplankton. *Marine and Freshwater Behaviour and Physiology*, **39(1)**, 13–24.

Naylor, E. and Atkinson, R. J. A. (1972). Pressure and the rhythmic behaviour of inshore animals. *Symposia of the Society for Experimental Biology*, **25**, 395–415.

Naylor, E. and Atkinson, R. J. A. (1976). Rhythmic behaviour of *Nephrops* and some other marine crustaceans. In *Perspectives in Experimental Biology*, **I**, ed. P. Spencer Davies. Oxford: Pergamon Press, pp. 135–143.

Naylor, E. and Isaac, M. J. (1973). Behavioural significance of pressure responses in megalopa larvae of *Callinectes sapidus* and *Macropipus* sp. *Marine Behaviour and Physiology*, **1**, 341–350.

Naylor, E. and Williams, B. G. (1968). Effects of eyestalk removal on rhythmic locomotor activity in *Carcinus*. *Journal of Experimental Biology*, **49**, 107–116.

Naylor, E. and Williams, B. G. (1984). Phase-responsiveness of the circatidal locomotor activity rhythm of *Hemigrapsus edwardsii* (Hilgendorf) to high tide. *Journal of the Marine Biological Association UK*, **64**, 81–90.

Naylor, E., Smith, G. and Williams, B. G. (1971). The role of the eyestalk in the tidal activity rhythm of the shore crab *Carcinus maenas* (L.). In *Neurobiology of Invertebrates*, ed. J. Salánki. Budapest: Akadémiai Kiadó, pp. 423–429.

Neil, W. E. (1992). Population variation in the ontogeny of predator-induced vertical migration of copepods. *Nature*, **356**, 54–57.

Nelson, J. (1912). Report of the Biological Department of the New Jersey Agricultural Experiment Station for the year 1911.

Neumann, D. (1965). Photoperiodische steuerung der 15-tagigen lunaren metamorphose Periodik von *Clunio* population (Diptera, Chironomidae). *Zeitschrift für Naturforschung*, **206**, 818–819.

Neumann, D. (1976a). Entrainment of a semi-lunar rhythm by simulated cycles of mechanical disturbance. *Journal of Experimental Marine Biology and Ecology*, **35**, 73–85.

Neumann, D. (1976b). Entrainment of a semilunar rhythm. In *Biological Rhythms in the Marine Environment*, ed. P. J. DeCoursey. Columbia, SC, USA: University of South Carolina Press, pp. 115–127.

Neumann, D. (1981). Tidal and lunar rhythms. In *Handbook of Behavioural Neurobiology*, **4**, ed. J. Aschoff. New York: Plenum Press, pp. 351–380.

Neumann, D. (1987). Tidal and lunar adaptations of reproductive activities in invertebrate species. In *Comparative Physiology of Environmental Adaptations*, **III**, ed. L. Pevet. Basel: Karger, pp. 152–170.

Nichols, J. H., Thompson, B. M. and Cryer, M. (1982). Production, drift and mortality of the planktonic larvae of the edible crab (*Cancer pagurus*) off the north east coast of England. *Netherlands Journal of Sea Research*, **16**, 173–184.

Njus, D., Sulzman, F. and Hastings, J. W. (1974). Membrane model for the circadian clock. *Nature*, **248**, 116–120.

Northcott, S. J., Gibson, R. N. and Morgan, E. (1990). The persistence and modulation of endogenous circatidal rhythmicity in *Lipophrys pholis* (Teleostei). *Journal of the Marine Biological Association UK*, **70**, 815–827.

Northcott, S. J., Gibson, R. N. and Morgan, E. (1991). Phase-resposiveness and modulation of endogenous circatidal rhythmicity in *Lipophrys pholis* (Teleostei). *Journal of Experimental Marine Biology and Ecology*, **148**, 47–57.

Nultsch, W., Rüffer, U. and Pfau, J. (1984). Circadian rhythms in chromatophore movements of *Dictyota dichotoma*. *Marine Biology*, **81**, 217–222.

Olive, P. J. W., Lewis, C. and Beardall, V. (2000). Fitness components of seasonal reproduction: an analysis using *Nereis virens* as a life history model. *Oceanologica Acta*, **23**, 377–389.

Ouyang, Y., Anderson, C. R., Kondo, T., Golden, S. S. and Johnson, C. H. (1998). Resonating circadian clocks enhance fitness in cyanobacteria. *Proceedings of the National Academy of Sciences, Washington*, **951**, 8660–8664.

Page, T. L. and Barrett, R. K. (1987). Effects of light on pacemaker development. *Journal of Comparative Physiology A*, **165**, 51–59.

Page, T. L. and Larimer, J. L. (1972). Entrainment of the circadian locomotor activity rhythm in crayfish. *Journal of Comparative Physiology*, **78**, 107–120.

Page, T. L. and Larimer, J. L. (1975a). Neural control of circadian rhythmicity in the crayfish: I. The locomotor rhythm. *Journal of Comparative Physiology*, **97**, 59–80.

Page, T. L. and Larimer, J. L. (1975b). Neural control of circadian rhythm in crayfish: II. The ERG amplitude rhythm. *Journal of Comparative Physiology*, **97**, 81–96.

Palmer, J. D. (1974). *Biological Clocks in Marine Animals*. New York/London: Wiley and Sons, 173 pp.

Palmer, J. D. (1976). Clock-controlled vertical migration rhythms in intertidal organisms. In *Biological Rhythms in the Marine Environment*, ed. P. J. Decoursey. Columbia, SC, USA: University of South Carolina Press, pp. 239–255.

Palmer, J. D. (1995a). *The Biological Rhythms and Clocks of Intertidal Animals*. Oxford, UK: Oxford University Press, 217 pp.

Palmer, J. D. (1995b). Review of the dual clock control of tidal rhythms and the hypothesis that the same clock controls both circatidal and circadian rhythms. *Chronobiology International*, **12(5)**, 299–310.

Palmer, J. D. (1997). Duelling hypotheses: circatidal versus circalunidian battle basics. *Chronobiology International*, **14**, 337–346.

Palmer, J. D. and Round, F. E. (1965). Persistent, vertical-migration rhythms in benthic microflora: I. The effect of light and temperature on the rhythmic behaviour of *Euglena obtusata*. *Journal of the Marine Biological Association UK*, **45**, 567–582.

Pape, C. and Lüning, K. (2006). Quantification of melatonin in phototrophic organisms. *Journal of Pineal Research*, **41**, 157–165.

Papi, F. (2006). Navigation of marine, freshwater and coastal animals: concepts and current problems. *Marine and Freshwater Behaviour and Physiology*, **39**, 3–12.

Papi, F. and Pardi, L. (1953). Ricerche sull'orientamento di *Talitrus saltator* (Montagu)(Crustacea, Amphipoda). II. *Zeitschrifte fur vergleichende Physiologie*, **35**, 430–518.

Papi, F., Luschi, P., Åkesson, S., Capogrossi, S. and Hays, G. C. (2000). Open-sea migration of magnetically disturbed sea turtles. *Journal of Experimental Biology*, **203**, 3435–3443.

Pardi, L. and Ercolini, A. (1986). Zonal recovery mechanisms in talitrid crustaceans. *Bollettino di Zoologia*, **53**, 139–160.

Pardi, L. and Papi, F. (1953). Ricerche sull'orientamento di *Talitrus saltator* (Montagu)(Crustacea, Amphipoda). I. *Zeitschrift fur vergleischende Physiologie*, **35**, 459–489.

Perkins, E. J. (1974). *The Biology of Estuaries and Coastal Waters*. London: Academic Press, 678 pp.

Permata, W. D., Kinzi, R. A. and Hidaka, M. (2000). Histological studies on the origin of planulae of the coral *Pocillopora damicornis*. *Marine Ecology Progress Series*, **200**, 191–200.

Petpiroon, S. and Morgan, E. (1983). Observations on the tidal activity rhythm of the periwinkle *Littorina nigrolineata* (Gray). *Marine Behaviour and Physiology*, **9**, 171–192.

Phillips, B. F. (1977). A review of the larval ecology of rock lobsters. In *Workshop on Lobster and Rock Lobster Ecology and Physiology*, **7**, eds. B. F. Phillips and J. S. Cobb. Australia: CSIRO, pp. 175–185.

Phillips, B. F. (1981). The circulation of the southeastern Indian Ocean and the planktonic life of the western rock lobster. *Oceanography and Marine Biology Annual Reviews*, **19**, 11–39.

Pineda, J. (1994). Internal tidal bores in the nearshore: warm-water fronts, seaward gravity currents and onshore transport of neustonic larvae. *Journal of Marine Research*, **52**, 427–458.

Pittendrigh, C. S. (1960). Circadian rhythms and the circadian organization of living systems. *Cold Spring Harbor Symposia on Quantitative Biology*, **25**, 159–184.

Pittendrigh, C. S. (1961). On temporal organization in living systems. *Harvey Lecture Series*, **56**, 93–125.

Pittendrigh, C. S. (1993). Temporal organization: reflections of a Darwinian clock-watcher. *Annual Review of Physiology*, **55**, 17–54.

Pittendrigh, C. S. and Minnis, D. H. (1972). Circadian systems: longevity as a function of circadian resonance in *Drosophila melanogaster*. *Proceedings of the National Academy of Science USA*, **69**, 1537–1539.

Prince, P. A., Wood, A. G., Barton, T. and Croxall, J. P. (1992). Satellite tracking of wandering albatrosses (*Diomedea exulans*) in the south Atlantic. *Antarctic Science*, **4**, 31–36.

Pugh, D. T. (1987). *Tides, Surges and Mean Sea-level*. Chichester, UK: J. Wiley and Sons, 472 pp.

Quinn, T. P. and Brannon, E. L. (1982). The use of celestial and magnetic cues by orienting sockeye salmon smolts. *Journal of Comparative Physiology*, **147**, 547–552.

Rajan, K. P., Kharouf, H. H. and Lockwood, A. P. M. (1979). Rhythmic cycles of blood sugar concentrations in the crab *Carcinus maenas*. *European Marine Biology Symposia*, **13**, 415–422.

Ralph, M. R., Foster, R. G., Davis, F. L. and Menaker, M. (1990). Transplanted suprachiasmatic nucleus determines circadian period. *Science*, **247**, 975–978.

Reid, D. G. (1988). The diurnal modulation of the circatidal activity rhythm by feeding in the isopod *Eurydice pulchra*. *Marine Behaviour and Physiology*, **6**, 273–285.

Reid, D. G. and Naylor, E. (1985). Free-running, endogenous semilunar rhythmicity in a marine isopod crustacean. *Journal of the Marine Biological Association UK*, **65**, 85–91.

Reid, D. G. and Naylor, E. (1986). An entrainment model for semilunar rhythmic swimming behaviour in the marine isopod *Eurydice pulchra* Leach. *Journal of Experimental Marine Biology and Ecology*, **100**, 25–35.

Reid, D. G. and Naylor, E. (1989). Are there separate circatidal and circadian clocks in the shore crab *Carcinus maenas*? *Marine Ecology Progress Series*, **52**, 1–6.

Reid, D. G. and Naylor, E. (1990). Entrainment of bi-modal circatidal rhythms in the shore crab *Carcinus maenas*. *Journal of Biological Rhythms*, **5**, 333–447.

Reid, D. G., Warman, C. G. and Naylor, E. (1992). Ontogenetic changes in zeitgeber action in the tidally rhythmic behaviour of the shore crab *Carcinus maenas* (L.). *European Marine Biological Symposia*, **27**, 129–133.

Richardson, C. A. (1987). Microgrowth patterns in the shell of the Malaysian cockle *Anadara granosa* (L.) and their use in age determination. *Journal of Experimental Marine Biology and Ecology*, **111**, 77–98.

Richardson, C. A. (1989). An analysis of the microgrowth bands in the shell of the common mussel *Mytilus edulis*. *Journal of the Marine Biological Association UK*, **69**, 477–491.

Rimmer, D. W. and Phillips, B. F. (1979). Diurnal migration and vertical distribution of phyllosoma larvae of the western rock lobster *Palinurus cygnus*. *Marine Biology*, **54**, 109–124.

Ritz, D. A. (1994). Social aggregation in pelagic invertebrates. *Advances in Marine Biology*, **30**, 155–216.

Rodriguez, G. and Naylor, E. (1972). Behavioural rhythms in littoral prawns. *Journal of the Marine Biological Association UK*, **52**, 81–95.

Roe, H. S. J. (1974). Observations on the diurnal vertical migrations of an oceanic animal community. *Marine Biology*, **28**, 99–113.

Roe, H. S. J. (1983). Vertical distribution of euphausiids and fish in relation to light intensity in the north eastern Atlantic. *Marine Biology*, **77**, 287–298.

Roe, H. S. J. (1984). The diel migrations and distributions within a mesopelagic community in the north east Atlantic. 2. Vertical migrations and feeding of mysids and decapod Crustacea. *Progress in Oceanography*, **13**, 269–318.

Roenneberg, T. and Morse, D. (1993). Two circadian oscillators in one cell. *Nature*, **362**, 362–364.

Rose, M. (1925). Contributiona l'etude de la Biologique du Plankton. Le probleme des migrations verticals journalieres. *Archives de Zoologie Experimentale et Generale*, **64**, 387–542.

Round, F. E. and Palmer, J. D. (1966). Persistent vertical migration rhythms in benthic microflora. II. Field and laboratory studies of diatoms from the banks of the River Avon. *Journal of the Marine Biological Association UK*, **46**, 191–214.

Russell, F. S. (1925). The vertical distribution of marine macro-plankton: an observation on diurnal changes. *Journal of the Marine Biological Association UK*, **13**, 769–809.

Russell, F. S. (1927). The vertical distribution of plankton in the sea. *Biological Reviews*, **2(3)**, 213–262.

Ryland, J. S. (2000). European marine biology: past, present and future. *Biologia Marina Mediterranea*, **7**, 1–27.

Saigusa, M. (1980). Entrainment of a semilunar rhythm by simulated moonlight cycle on the terrestrial crab, *Sesarma haematocheir*. *Oecologia*, **46**, 38–44.

Saigusa, M. (1982). Larval release rhythm coinciding with solar day and tidal cycles in the terrestrial crab *Sesarma*; harmony with the semilunar timing and its adaptive significance. *Biological Bulletin*, **162**, 371–386.

Saigusa, M. (1986). The circatidal rhythm of larval release in the incubating crab, *Sesarma*. *Journal of Comparative Physiology*, **159**, 21–31.

Sato, M. and Jumars, P. A. (2008). Seasonal and vertical variations in emergence behaviour of *Neomysis americana*. *Limnology and Oceanography*, **53**(4), 1665–1677.

Saunders, D. S. (1972). Circadian control of larval growth in *Sarcophaga argyrostoma*. *Proceedings of the National Academy of Sciences USA*, **69**, 2738–2740.

Saunders, D. S. (1982). Photoperiodism in animals and plants. In *Biological Timekeeping*, ed. J. Brady. *Society for Experimental Biology Seminar Series*, **14**, 65–82.

Scapini, F. (1986). Inheritance of solar direction-finding in sandhoppers. 4. Variation in the accuracy of orientation with age. *Monitore Zoologia Italiano*, **20**, 53–61.

Scapini, F. and Buiatti, M. (1985). Inheritance of solar direction-finding in sandhoppers. *Journal of Comparative Physiology A*, **157**, 433–440.

Scapini, F., Rossano, C., Marchetti, G. M. and Morgan, E. (2005). The role of the biological clock in the sun–compass orientation of free-running individuals of *Talitrus saltator*. *Animal Behaviour*, **69**(4), 835–843.

Schaffelke, B. and Lüning, K. (1994). A circannual rhythm controls seasonal growth in the kelps *Laminaria hyperborea* and *L. digitata* from Helgoland (North Sea). *European Journal of Phycology*, **29**, 49–56.

Scheltema, R. S. (1971). The dispersal of the larvae of shoal-water benthic invertebrates over long distances by ocean currents. *European Marine Biology Symposia*, **4**, 7–28.

Scheltema, R. S. (1975). Relationship of larval dispersal, gene-flow and natural selection to geographical variation of benthic invertebrates in estuaries and along coastal regions. In *Estuarine Research*, **I**, ed. L. E. Cronin. New York: Academic Press, pp. 373–391.

Schweiger, H. G. and Schweiger, M. (1977). Circadian rhythms in unicellular organisms: an endeavour to explain the molecular mechanism. *International Review of Cytology*, **51**, 315–342.

Schmidt-Koenig, K. (1975). *Avian Orientation and Navigation*. London: Academic Press, 180 pp.

Sehgal, A., Rothenfluh-Hilfiker, A., Hunter-Ensor, M. *et al.* (1995). Rhythmic expression of *timeless*: a basis for promoting circadian cycles in *period* gene autoregulation. Science, **270**, 808–810.

Serfling, S. A. and Ford, R. F. (1975). Ecological studies of the puerulus larval stage of the Californian spiny lobster *Panulirus interruptus*. *National Oceanic and Atmospheric Administration (USA), Fisheries Bulletin*, **73**, 360–367.

Shanks, A. L. (1995). Mechanisms of cross-shelf dispersal of larval invertebrates and fish. In *Ecology of Marine Invertebrate Larvae*, ed. L. R. McEdward. Boca Raton, Florida: CRC, pp. 323–368.

Shepard, E. L. C., Ahmed, M. Z., Southall, E. J. *et al.* (2006). Diel and tidal rhythms in diving behaviour of pelagic sharks identified by signal processing of archival tagged data. *Marine Ecology Progress Series*, **328**, 205–213.

Siegal, D. A., Kinlan, B. P., Gaylord, B. and Gaines, S. D. (2003). Lagrangian descriptions of marine larval dispersion. *Marine Ecology Progress Series*, **260**, 83–96.

Singarajah, K. V., Moyse, J. and Knight-Jones, E. W. (1967). The effect of feeding upon the phototactic behaviour of cirripede nauplii. *Journal of Experimental Marine Biology and Ecology*, **1**, 144–153.

Skov, M. W., Hartnoll, R. G., Ruwa, R. K. *et al.* (2005). Marching to a different drummer: crabs synchronize reproduction to a 14-month lunar-tidal cycle. *Ecology*, **86(5)**, 1164–1171.

Skud, B. E. (1967). Responses of marine organisms during the solar eclipse of July 1963. *US Fish and Wildlife Service Fishery Bulletin*, **66**, 259–271.

Smith, G. and Naylor, E. (1972). The neurosecretory system of the eyestalk of *Carcinus maenas* (Crustacea: Decapoda). *Journal of Zoology, London*, **166**, 313–321.

Smith, T. J. (1990). Phylogenetic distribution and function of arylalkylamine N-acetyltransferase. *Bioassays*, **12**, 30–33.

Southward, A. J. and Crisp, D. J. (1965). Activity rhythms in barnacles in relation to respiration and feeding. *Journal of the Marine Biological Association UK*, **45**, 161–185.

Sommer, H. H. (1972). Endogene und exogene periodik in der aktivitat eines medeeren krebses (*Balanus balanus* L.). *Zeitschrift fur vergleichende Physiologie*, **76**, 177–192.

Sponaugle, S. and Pinkard, D. (2004). Lunar cyclic population replenishment of a coral reef fish: shifting patterns following oceanic events. *Marine Ecology Progress Series*, **267**, 267–280.

Steele, J. H. (1976). Patchiness. In *The Ecology of the Seas*, eds. D. H. Cushing and J. J. Walsh. Oxford, UK: Blackwell, pp. 98–115.

Stephan, F. K. and Nunez, A. A. (1977). Elimination of circadian rhythms in drinking, activity, sleep and temperature by isolation of the suprachiasmatic nuclei. *Behavioral Biology*, **20**, 1–16.

Strumwasser, F. (1965). The demonstration and manipulation of a circadian rhythm in a single neuron. In *Circadian Clocks*, ed. J. Aschoff. Amsterdam: North Holland Publishing Co., pp. 442–462.

Sulkin, S. D. (1984). Behavioural basis of depth regulation in the larvae of brachyuran crabs. *Marine Ecology Progress Series*, **15**, 181–205.

Sulkin, S. D., Van Heukelem, W. F., Kelly, P and Van Heukelem, L. (1980). The behavioural basis of larval recruitment in the crab *Callinectes sapidus* Rathbun: a laboratory investigation of ontogenetic changes in geotaxis and barokinesis. *Biological Bulletin*, **159**, 402–417.

Suzuki, L. and Johnson, C. H. (2001). Algae know the time of day: circadian and photoperiodic programs. *Journal of Phycology*, **37**, 933–942.

Sweeney, B. M. (1969a). *Rhythmic Phenomena in Plants*. London/New York: Academic Press.

Sweeney, B. M. (1969b). Circadian rhythms in plants. In *Physiology of Plant Growth and Development*, ed. M. B. Wilkins. London: McGraw Hill, pp. 647–671.

Sweeney, B. M. (1976). Pros and cons of the membrane model for circadian rhythms in the marine algae *Gonyaulax* and *Acetabularia*. In *Biological Rhythms in the Marine Environment*, ed. P. J. DeCoursey. Columbia, SC, USA: University of South Carolina Press, pp. 63–76.

Sweeney, B. M. and Borgese, M. B. (1989). A circadian rhythm in cell division in a prokaryote, the cyanobacterium *Synechococcus* WH 7803. *Journal of Phycology*, **25**, 183–186.

Sweeney, B. M. and Hastings, J. W. (1957). Characteristics of the diurnal rhythm of luminescence in *Gonyaulax polyedra*. *Journal of Cellular and Comparative Physiology*, **49**, 115–128.

Sweeney, B. M. and Haxo, F. T. (1961). Persistence of a photosynthetic rhythm in enucleated *Acetabularia*. *Science*, **134**, 1361–1363.

Sweeney, B. M., Tuffli, C. F. and Rubin, R. H. (1967). The circadian rhythm of photosynthesis in *Acetabularia* in the presence of actinomycin D, puromycin and chloramphenicol. *Journal of General Physiology*, **50**, 647–659.

Tankersley, R. A., McKelvey, L. M. and Forward, R. B. (1995). Responses of estuarine crab megalopae to pressure, salinity and light: implications for flood-tide transport. *Marine Biology*, **122**, 391–400.

Tankersley, R. A., Welch, J. M. and Forward, R. B. (2002). Settlement time of blue crab (*Callinectes sapidus*) megalopae during flood-tide transport. *Marine Biology*, **141**, 863–875.

Taylor, A. C. and Naylor, E. (1977). Entrainment of the locomotor rhythm of *Carcinus* by cycles of salinity change. *Journal of the Marine Biological Association UK*, **57**, 273–277.

Tester, P. A., Cohen, J. H. and Cervetto, G. (2004). Reverse vertical migration and hydrographic distribution of *Anomalocera ornata* (Copepoda: Pontellidae) in the US South Atlantic Bight. *Marine Ecology Progress Series*, **268**, 195–203.

Thomas, B. and Vince-Prue, D. (1997). *Photoperiodism in Plants*. San Diego: Academic Press, 428 pp.

Thomas, N. J., Lasiak, T. A. and Naylor, E. (1981). Salinity preference behaviour in *Carcinus*. *Marine Behaviour and Physiology*, **7**, 277–283.

Thompson, W. F. (1919). The spawning of the grunion *Leuresthes tenuis*. *Californian Fish and Game Committee Bulletin*, **3**, 1–29.

Thorson, G. (1964). Light as an ecological factor in the dispersal and settlement of larvae of marine bottom invertebrates. *Ophelia*, **1**, 167–208.

Thurman, C. L. (2004). Unravelling the ecological significance of endogenous rhythms in intertidal crabs. *Biological Rhythm Research*, **35**, 43–67.

Tilberg, C. F., Kernehan C. D., Andon, A. and Epifanio, C. E. (2008). Modeling estuarine ingress of blue crab megalopae: effects of temporal patterns of larval release. *Journal of Plankton Research*, **66**, 391–412.

Tilden, A., McGann, L., Schwartz, J., Bowe, A. and Salazar, C. (2001). Effect of melatonin on haemolymph glucose and lactate levels in the fiddler crab *Uca pugilator*. *Journal of Experimental Zoology* A, **290**, 379–383.

Tilden, A. R., Shanahan, J. K., Khilji, Z. S. *et al.* (2003). Melatonin and locomotor activity in the fiddler crab *Uca pugilator*. *Journal of Experimental Zoology* A, **297**, 80–87.

Titlyanov, E. A., Titlyanov, T. V. and Lüning, K. (1996). Diurnal and circadian periodicity of mitosis and growth in marine macroalgae II. The green alga *Ulva pseudocurvata*. *European Journal of Phycology*, **31**, 181–188.

Uglow, J. (2002). *The Lunar Men*. London: Faber, 501 pp.

Ugolini, A. (2003). Activity rhythms and orientation in sandhoppers (Crustacea: Amphipoda). *Frontiers in Biosciences*, **8**, 722–732.

Ugolini, A., Somigli, S. and Mercatelli, L. (2005a). Green land and blue sea: a coloured landscape in the orientation of the sandhopper *Talitrus saltator* (Montagu) (Amphipoda: Talitridae). *Journal of Experimental Biology*, **209**, 2509–2514.

Ugolini, A., Boddi, V., Mercatelli, L. and Castellini, C. (2005b). Moon orientation in adult and young sand hoppers under artificial light. *Proceedings of the Royal Society London, B*, **272**, 2189–2194.

Ugolini, A., Somigli, S., Pasquali, V. and Renzi, P. (2007). Locomotor activity rhythm and sun compass orientation in the sand hopper *Talitrus saltator* are related. *Journal of Comparative Physiology*, **193**, 1259–1263.

Van Tassel, D. L., Roberts, N., Lewy, A. and O'Neil, S. D. (2001). Melatonin in plant organs. *Journal of Pineal Research*, **31**, 8–15.

Vargas, C. A., Narvaez, D. A., Pinones, A., Venegas, R. M. and Navarrete, S. A. (2004). Internal tidal bores, warm fronts and settlement of invertebrates in central Chile. *Estuarine Coastal and Shelf Science*, **61**, 603–612.

Vielhaben, V. (1963). Zur Deutung des semilunaren Fortpflanzungszyklus von *Dictyota dichotoma*. *Zeitschrift fur Botanik*, **51**, 156–173.

Vince-Prue, D. (1982). Phytochrome and photoperiodic physiology in plants. In *Biological Timekeeping*, ed. J. Brady. *Society for Experimental Biology Seminar Series*, **14**, 101–117.

Vinogradov, M. E. (1968). Vertical distribution of the oceanic plankton. Moscow. (Translated by U.S. Department of Commerce, 1970), 339 pp.

Vitaterna, M. H., King, D. P., Chang, A. *et al.* (1994). Mutagenesis and mapping of a mouse gene, *clock*, essential for circadian behaviour. *Science*, **264**, 719–725.

Vivien-Roels, B. and Pevet, P. (1993). Melatonin: presence and formation in invertebrates. *Experientia*, **49**, 642–647.

Vivien-Roels, B., Pevet, P., Beck, O. and Fevre-Montagne, M. (1984). Identification of melatonin in the compound eyes of an insect, the locust (*Locusta migratoria*), by radioimmunoassay and gas chromatography-mass spectrometry. *Neuroscience Letters*, **49**, 153–157.

Ward, S. (1992). Evidence for broadcast spawning as well as brooding in the scleratinian coral *Pocillopora damicornis*. *Marine Biology*, **112**, 641–646.

Warman, C. G. and Naylor, E. (1995). Evidence for multiple, cue-specific circatidal clocks in the shore crab *Carcinus maenas*. *Journal of Experimental Marine Biology and Ecology*, **189**, 93–101.

Warman, C. G., O'Hare, T. J. and Naylor, E. (1991a). Vertical swimming in wave-induced currents as a control mechanism of intertidal migration by a sand-beach isopod. *Marine Biology*, **111**, 49–54.

Warman, C. G., Abello, P. and Naylor, E. (1991b). Behavioural responses of *Carcinus mediterraneus* Czerniavsky 1884 to changes in salinity. *Scientia Marina*, **54(4)**, 637–643.

Warman, C. G., Reid, D. G. and Naylor, E. (1993a). Variation in the tidal migratory behaviour and rhythms of light-responsiveness in the shore crab *Carcinus maenas*. *Journal of the Marine Biological Association UK*, **73**, 355–364.

Warman, C. G., Reid, D. G. and Naylor, E. (1993b). Circatidal variability in the behavioural responses of a sand-beach isopod *Eurydice pulchra* (Leach) to orientational cues. *Journal of Experimental Marine Biology and Ecology*, **168**, 59–70.

Watson, G. J., Williams, M. E. and Bentley, M. G. (2000). Can synchronous spawning be predicted from environmental parameters? A case study of the lugworm *Arenicola marina*. *Marine Biology*, **136**, 1003–1017.

Weinberg, S. (1999). *A Fish Caught in Time*. London: Fourth Estate, 241 pp.

Welch, J. M. and Forward, R. B. (2001). Flood tide transport of blue crab, *Callinectes sapidus*, postlarvae: behavioural responses to salinity and turbulence. *Marine Biology*, **139**, 911–918.

Welch, J. M., Forward, R. B. and Howd, P. A. (1999). Behavioural responses of blue crab *Callinectes sapidus* postlarvae to turbulence: implications for selective tidal stream transport. *Marine Ecology Progress Series*, **179**, 135–143.

Welsh, D. K., Logothetis, D. E., Meister, M. and Reppert, S. M. (1995). Individual neurons dissociated from rat suprachiasmatic nucleus express independently-phased circadian firing rhythms. *Neuron*, **14**, 697–706.

Wheeler, D. E. (1978). Semilunar hatching periodicity in the mud fiddler crab *Uca pugnax*. *Estuaries*, **1**, 268–269.

White, R. G., Hill, A. E. and Jones, D. A. (1988). Distribution of *Nephrops norvegicus* (L.) larvae in the western Irish Sea: an example of advective control on recruitment. *Journal of Plankton Research*, **10**(4), 735–747.

Wilcockson, D. C. and Zhang, L. (2008). Circatidal clocks. *Current Biology*, **18**(17), R753.

Wilcockson, D. C., Chung, J. S. and Webster, S. G. (2002). Is crustacean hyperglycaemic hormone precursor-related peptide a circulating neurohormone? *Cell and Tissue Research*, **307**, 129–138.

Williams, B. G. (1998). The lack of circadian timing in two intertidal invertebrates and its significance in the circatidal/circalunidian debate. *Chronobiology International*, **15**, 205–208.

Williams, B. G. and Naylor, E. (1967). Spontaneously induced rhythm of tidal periodicity in laboratory-reared *Carcinus*. *Journal of Experimental Biology*, **47**, 229–234.

Williams, B. G. and Naylor, E. (1969). Synchronization of the locomotor rhythm of *Carcinus*. *Journal of Experimental Biology*, **51**, 715–725.

Williams, B. G., Naylor, E. and Chatterton, T. D. (1985). The activity patterns of New Zealand mud crabs under field and laboratory conditions. *Journal of Experimental Marine Biology and Ecology*, **89**, 269–282.

Williams, J. A. (1979). A semilunar rhythm of locomotor activity and moult synchrony in the sand-beach amphipod *Talitrus saltator*. *European Marine Biology Symposia*, **13**, 407–414.

Williams, J. A. (1980). The light-response rhythm and seasonal entrainment of the endogenous circadian locomotor rhythm of *Talitrus saltator* (Crustacea: Amphipoda). *Journal of the Marine Biological Association UK*, **60**, 773–785.

Williams, J. A., Pullin, R. S. V., Williams, B. G., Arechiga, H. and Naylor, E. (1979a). Evaluation of the effects of injected eyestalk extract on rhythmic locomotor activity in *Carcinus*. *Comparative Biochemistry and Physiology*, **62**, 903–907.

Williams, J. A., Pullin, R. S. V., Naylor, E., Smith, G. and Williams, B. G. (1979b). The role of Hanstrom's Organ in clock control in *Carcinus maenas*. *European Marine Biology Symposia*, **13**, 459–466.

Williams, Julie A. and Sehgal, A. (2001). Molecular components of the circadian system in *Drosophila*. *Annual Review of Physiology*, **63**, 729–755.

Williams, J. L. (1898). Reproduction in *Dictyota dichotoma*. *Annals of Botany*, **12**, 559–560.

Williams, J. L. (1905). Studies on the Dictyotaceae III. The periodicity of the sexual cells of *Dictyota dichotoma*. *Annals of Botany*, **19**, 531–560.

Williamson, D. I. (1951). Studies on the biology of Talitridae (Crustacea; Amphipod): visual orientation in *Talitrus saltator*. *Journal of the Marine Biological Association UK*, **30**, 91–99.

Winfree, A. T. (1987). *The Timing of Biological Clocks*. Scientific American Books, 200 pp.

Withyachumnarnkul, B., Pongsa-Asawapaiboon, A., Ajpru, S. *et al.* (1992). Continuous light increases N-acetyltransferase activity in the optic lobe of the giant freshwater prawn *Macrobrachium rosenbergii* de Man (Crustacea, Decapoda). *Life Sciences*, **51**, 1479–1484.

Wood, L. and Hargis, J. H. (1971). Transport of bivalve larvae in a tidal estuary. *European Marine Biology Symposia*, **4**, 29–44.

Wooldridge, T. and Erasmus, T. (1980). Utilization of tidal currents by estuarine plankton. *Estuarine, Coastal and Shelf Science*, **11**, 107–114.

Zanchin, G. (2001). Macro- and microcosmus: moon influence on the human body. *Earth, Moon and Planets*, **85–86**, 453–461.

Zeng, C. and Naylor, E. (1996a). Synchronization of endogenous tidal vertical migration rhythms in laboratory-hatched larvae of the crab *Carcinus maenas*. *Journal of Experimental Marine Biology and Ecology*, **198**, 269–289.

Zeng, C. and Naylor, E. (1996b). Endogenous tidal rhythms of vertical migration in field-collected zoea-1 larvae of the shore crab *Carcinus maenas*: implications for ebb tide offshore dispersal. *Marine Ecology Progress Series*, **132**, 71–82.

Zeng, C. and Naylor, E. (1996c). Heritability of circatidal vertical migration rhythms in zoea larvae of the crab *Carcinus maenas* (L.). *Journal of Experimental Marine Biology and Ecology*, **202**, 239–257.

Zeng, C. and Naylor, E. (1997). Rhythms of larval release in the shore crab *Carcinus maenas* (Decapoda: Brachyura). *Journal of the Marine Biological Association UK*, **77**, 299–305.

Zeng, C., Abello, P. and Naylor, E. (1999). Endogenous tidal and semilunar moulting rhythms in early juvenile shore crabs *Carcinus maenas*: implications for adaptation to a high intertidal habitat. *Marine Ecology Progress Series*, **191**, 257–266.

Author index

Subject index